Dessin N° 5719
D. 8. 120. Delage
Maurice Delage

Delage

Styling and Design

La Belle Voiture Française

Richard S. Adatto and Diana E. Meredith

Notes and Acknowledgments

Delage Styling and Design - La Belle Voiture Française is a celebration of the beautiful French cars of the golden era between 1929 and 1953, with a special emphasis on the years just before World War II. With old and new photos, lovely period drawings and sketches and unpublished documents, we pay tribute to the styling and design of the cars created during this time. Although we recognize that the racing cars produced the technical platform for the ensuing luxury cars, and that racing put the marque on the map, this book is not a detailed history of the evolution of the company or its racing triumphs, which we leave to other enthusiasts. We include an appendix with abundant technical information for the avid tinkerer and history buff.

Friends and colleagues have provided help and support in producing this book. They have lent us documents and photos, corrected our errors, provided advice and inspired us to do better. Our thanks go to Margery Aronson, Susan Bendrick, Carolyn Brown, Jean-Paul Caron, Patrick Delage, Peter Jacobs, Sandra Kasky, Sam Mann, Glen Mounger, and Peter and Merle Mullin.

We have made every effort to be complete and accurate. Nearly every stone has been upturned to conduct the research required. Now we ask the reader to help complete the task by sending us your comments to contact@adattoauto.com, or via fax to +1-206-652-5255. We hope to add your material to subsequent editions and related projects, and will give full credit.

Richard S. Adatto and Diana E. Meredith

Delage

Styling and Design

La Belle Voiture Française

Richard S. Adatto and Diana E. Meredith

First published 2005

ISBN Casebound: 1-85443-204-4

Printed and bound in
Singapore by Star Standard

for the publisher

Dalton Watson Fine Books
1 Arundel Court, Elverlands Close,
Ferring, West Sussex BN12 5QE,
England

1730 Christopher Drive,
Deerfield, IL 60015,
USA

www.daltonwatson.com

DALTON WATSON FINE BOOKS

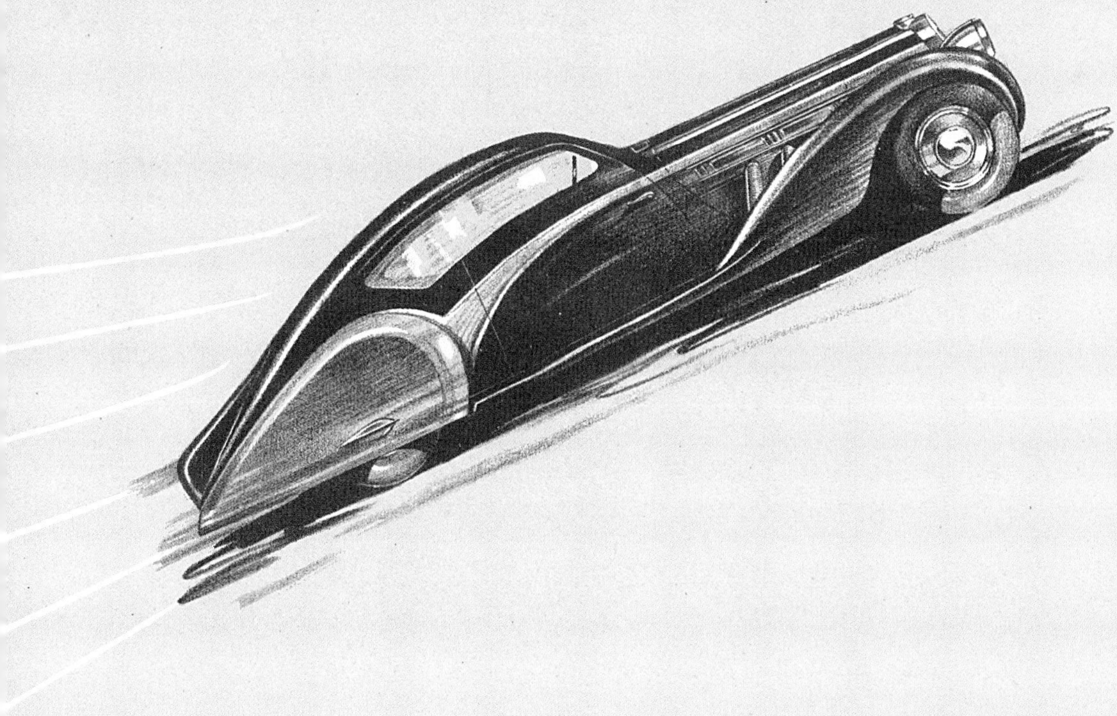

DELAGE

Contents

Introduction

Here at last we have a book in English that celebrates the styling and design of Louis Delage's beautiful French car, at that special moment when the superb Delage chassis came together with the best coachbuilders of the time.

Before the invention of the monocoque, a chassis was produced and sold separately from the body. The customer started off by selecting a bare chassis that then arrived at the coachbuilder's shop, where client and designer sat down to define what the dressed car would look like: a comfortable closed coupé, a sporty cabriolet or a racecar. Of the resulting chassis and coachbuilder combinations, I am fortunate to own a D8-120 Chapron cabriolet, which handles smoothly and easily, with plenty of power pulling steadily through all gears up to 100 mph plus. Driving this car through the Malibu Hills overlooking the Pacific Ocean, I think of Gene Kelly in the film *An American in Paris* sitting in the back seat of this same car, riding around Paris in high style and trying to woo Leslie Caron. No wonder Louis Delage is said to have quipped, "One drives an Alfa Romeo, one is driven in a Rolls, but one gives a Delage to one's mistress."

The handling of my Delage D6 Grand Prix car, a true racing thoroughbred, is entirely different. This 1946 factory-bodied model is very fast and lightweight. It sticks like glue to the road even at high speeds, but you can stand on the brakes and still hold a straight line. This probably accounts for its many successes in the post war years in Europe, including second place at Le Mans in 1949. In fact, in 1950 this car was the last Delage to race at Le Mans. Gear changing is a dream with the flick of the left hand at the steering column through the electro-magnetic Cotal gearbox.

Today any person lucky enough to own a Delage has an obligation to provide careful stewardship of this automobile, held only in trust for generations to come. Given their extraordinary engineering and exquisite styling, it is no surprise that these cars hold places of honor in museums and private collections throughout the world. We should care for them and respect them as true pieces of art.

Peter W. Mullin

0,88

Chapter I

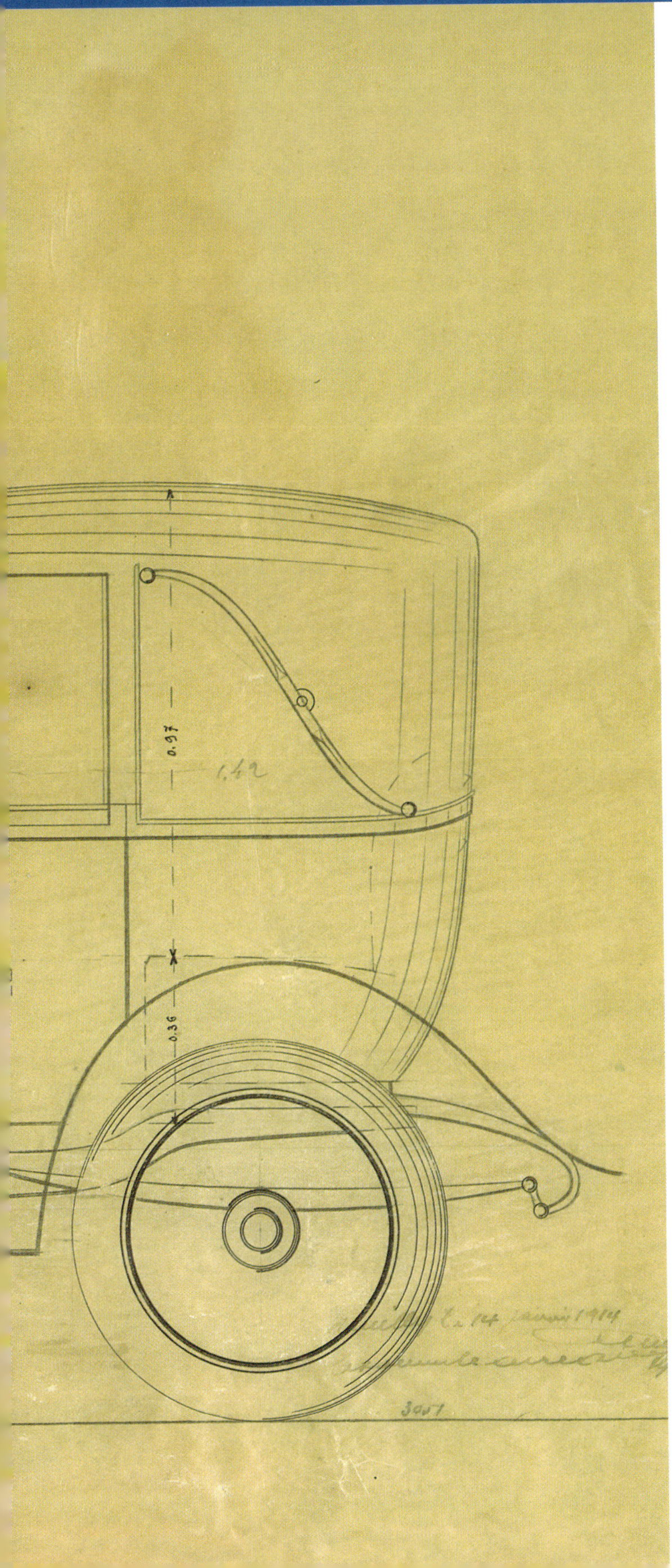

The Delage Company – Early Years

La Belle Voiture Française ... The Beautiful French Car

These simple words were coined by the public to describe the automobiles created by Louis Delage. But they represent something quite complex: admiration for the triumphs of Delage on the racetrack, recognition of the marque's technical contributions, appreciation for the glorious coachwork that was married to its chassis, and an overwhelming Gallic pride in a piece of art that came to define "French style" in the 1930s.

Louis Delage took the words to heart. "This very flattering description," he said, "was given to us by the public, which understood the essence of what our cars were all about and year after year imposed upon us the absolute obligation to become better and better." Thus, "The Beautiful French Car" became a slogan for one of France's oldest and most elegant automobile companies.

One hundred years after the foundation of the Delage Company, "The Beautiful French Car" remains entirely apt and a quick glance at the magnificent Delages presented in this book will convince you that you are looking at rolling art.

Louis Delage graduated from the École Nationale d'Arts et Métiers at Angers in 1893, completed his military service, and went to Paris, the hub of the automobile industry in France and a center of art and design in Europe. In 1900 he briefly set up shop as a consulting engineer, and then in 1903 joined Peugeot. These were early days for Peugeot, which had just moved from bicycles to cars, and provided an excellent environment for Delage to learn his craft and first imagine which direction his own company might take. In 1905, at the age of 31, he managed to obtain financial backing to start his own company and took with him the talented Augustin Legros, an engineer with whom he had worked at Peugeot. Legros became second in command to Delage, and although they often fought, disagreed, argued, and occasionally parted ways, they went on to design and produce great cars together for 31 years.

Starting in an old barn they developed two prototypes for the Paris Salon of December 1905. These were single-cylinder "voiturettes" (light cars, under 1.5-liters) with De Dion Bouton engines, and the public and press received them with an appreciation that turned into orders and money in the bank. As luck would have it, a wealthy investor visited the show and gave the Delage Company a loan - provided his son was hired - which meant there was enough capital to fund and create a new line.

Racing was a passion for Louis Delage and it led to great technological advances in his cars. It also brought him fame but only fleeting fortune, and in the long run it contributed to the company's bankruptcy. The first races were conducted over unpaved public roads, originally wagon tracks, and drew thousands of people who watched the cars drive by. Magazines and newspapers sponsored many of the early races and exhibitions and Delage soon understood the effect this could have on his marketing. He quickly entered two cars in the Coup des Voiturettes de l'Auto in Paris in 1906. A Sizaire et Naudin eventually won the race, and one of the Delages crashed into a tree, but the other placed second, beating the Peugeot. The crowd was impressed and the publicity generated among buyers and dealers greatly favored Delage. At the Paris Auto Salon of 1906, orders for cars poured in. Businessmen stood in line to become dealers and the company was on its way.

Pages 10-11 1914 styling study for Delage by Janssen & Cie., Paris.

Above The blue oval of the Delage Company.

Right "La Belle Voiture Française" - a sales piece by Louis Delage with English translation. According to Griffith Borgeson, writing in Automobile Quarterly (1976), Louis Delage's name "... is correctly spelled with a circumflex accent: Delâge. Louis Delage himself chose to drop the accent from his public use of the name, from his trademark, and even from his personal letterheads. He did use it, however, for his signature and, I am told, for his calling cards." In this book the accent is omitted.

"La Belle Voiture Française"

l'appelation, si flatteuse pour nous, par laquelle le public synthétise nos voitures, nous crée d'années en années, l'impérieux devoir de faire, constamment, de mieux en mieux.

Nos modèles, par la beauté de leurs lignes, l'agrément de leur conduite, leur confort, la sécurité et l'économie de leur utilisation la justifient plus encore que leurs aînés.

Leurs carrosseries, que ce soient celles de nos 16 cv. 6 cyl. ou celles de, la belle 25 CV huit cylindres, d'une ligne surbaissée impeccable, sont larges et spacieuses.

La gamme de nos modèles classiques se complète, naturellement, pour chacun d'eux, d'un modèle digne héritier de notre glorieux passé sportif:

vifs, gais, de belle allure et pleins d'entrain, "ce sont nos Types Sport"

Laissez-moi le bon espoir que vous accueillerez les uns et les autres avec autant de plaisir que nous avons pris de soin à les concevoir et à les réaliser.

L. Delâge.

"The Beautiful French Car"

This very flattering description was given to us by the public, which understands the essence of what our cars are all about and year after year imposes upon us the absolute obligation to become better and better.

The beauty of their line, the pleasure of their driving experience, their comfort, their safety and their reasonable upkeep, all make these models even more attractive than their predecessors.

The coachwork is wide and spacious in the 16 hp 6 cylinder models as well as in our beautiful 25 hp 8 cylinder model with its impeccable low slung design.

Each one of the classic models in this line is a worthy descendant of our glorious racing past.

Lively, fun, fast and full of spirit: These are our Sport Models.

I hope that you will welcome both types of cars with as much pleasure as we have experienced in their careful design and production.

L. Delâge

Above 1922 Delage CO town car. Coachbuilder is Arthur Boulogne.

Throughout the next several years the company grew, racing victories accumulated, and people the world over were enchanted by Delage. These triumphs turned into publicity, which in turn benefited the sale of touring cars. The marque came to be associated with great luxury and became an element of display for the wealthy, the sportsman and the elite.

In 1910 Louis Delage hired Arthur Michelat, another graduate from the École National d'Arts et Métiers. Legros became plant and marketing manager and Michelat was chief product designer and engineer. The team was later joined by engineer Maurice Gaultier. The factory expanded to a larger location at Boulevard de Courbevoie, and from the production line emerged a new chassis in 1913, the AH, with a Delage-designed six-cylinder 2.5-liter engine.

Once World War I started in 1914, the company - minus several of its key people - devoted itself to the war effort, producing ammunition, trucks and cars for the military. Remarkably, they managed to continue with the development of new cars, among them the BK Series 2, six-cylinder car in August 1915. This is considered to be the precursor of the post-war Type CO, with four wheel braking, twin carburetors and a six seat touring body. At the end of the war, Louis Delage was a very wealthy man. He bought his main residence on the Avenue Foch by the Bois de Boulogne and the Chateau de Pecq near the English Channel. He entertained in high style, sometimes hosting 500 guests at a time at his villa or on his new yacht, the *Oasis*. In addition, he bought a lavish house on the Champs-Elysées that he turned into a fashionable showroom from which to launch his Grand Touring luxury cars, which the public proclaimed "The Beautiful French Car." Sales were slow but better results came with the DE model created for the 1921 Paris Salon.

Over the next two years, 3,600 units of the DE line were sold, but then sales dropped and inventory piled up. In 1923 the company needed cash to operate and went public. Rather than use his new found funds to re-tool the factory, Louis Delage decided to build a "super car", determined to spare no expense in competing against Hispano-Suiza and Rolls-Royce with its "Best Car in the World" reputation. The resulting GL passenger car in 1924 was not a success, selling only 180 units.

The Delage Company entered into a difficult phase, partly due to the temperament of its leader - always the boss, always headstrong but not always right, and often unwilling to listen to others. During this time, some very talented engineers and designers were fired. The company finally stabilized in 1926 with the introduction of the DM model, which became its best seller.

Although sales of passenger cars were not always steady, on the racetrack these were great times. Delage racers participated in the European racing scene with some of the most sophisticated V-12 engines produced at the time. Reliable and fast, they won the Grand Prix de l'ACF in 1925 with the 2LCV and Robert Benoist at the wheel, and made Louis Delage a national hero. From this success he developed a straight-eight 1500 cc supercharged racecar which competed in the 1926-27 season, and in 1927 won all the major races in Europe, again with driver Robert Benoist. Once the 1927 season was over, and Delage's goal of winning the European Championship had been achieved, the team was disbanded. But the huge effort and funds devoted to racing had cost the company dearly.

Above A Delage presented at the 1925 Concours d'Elegance at the Bois de Boulogne, Paris. It is a sporty torpedo model with four seats.

57

Chapter II

Delage on the Racetrack

Rather than a complete racing history of Delage, this book provides an overview of the creative developments that turned into racing victories, making the company's name instantly recognizable among the public, and translating into sales of passenger and luxury cars.

From its establishment in 1905, the Delage Company was known for its many technical innovations. Funds were generally scarce, but Louis Delage had a particular knack for picking the best engineers, allowing them to spread their wings and develop many ingenious and modern advancements. He started off with a one-cylinder engine that he had to buy from De Dion, and ended by producing a world championship car in 1927, including his own Delage motor, a sophisticated, supercharged straight-eight, two overhead cam engine. The company ended its racing involvement after World War II with the D6, an inline six-cylinder race car that competed at Le Mans in 1949, coming to a second place finish behind a Ferrari 166MM driven by Luigi Chinetti and Lord Selsdon.

Louis Delage always understood that racing was an effective way to promote his new company. With only two employees in 1906 and very little capital he managed to have two racecars under construction during his first year of production. He chose the Voiturette class (small, light cars) for his first effort. Working with his engineer Augustin Legros, they based the cars on a single cylinder motor bought off the shelf from De Dion Bouton, to which they added their own chassis. In their first race, one Delage hit a tree and the next finished second, leaving the well-funded and prepared Peugeot team car in third place. The underdog had won and the public was delighted. The company was on its way.

The team entered the same race once more in 1907 with three cars, but its competitors had developed more horsepower for their motors and were not to be shown up again. Seventh place was the best the Delage racers could do, so Louis Delage decided to develop and build his own motors for the next season. With a lot of difficulty, and borrowing from motorcycle technology, they developed a sophisticated, four valve, two spark plug motor. This they placed on a new chassis for the Grand Prix des Voiturettes in 1908. They entered three cars and dominated the race again, relishing their role as "giant killers." The race was very popular with the public as it featured small and affordable cars. Racing fans were impressed, public reaction was favorable and the reputation of the company was now solidly established within the auto industry.

In 1909 the German Mercedes cars dominated racing, to the point that the French became incensed. In order to show their displeasure, they decided to boycott their own French races as well as the other racing events in Europe. Then came 1910, a breakthrough year for Delage. Arthur Michelat replaced Augustin Legros as chief engineer and designed a new and bigger car for the 1911 season. The Delage Type X was a very early twin cam, three-liter, four-cylinder motor, at the time a rare and expensive item in the United States. On July 25, 1911 in the Coupe de l'Auto - one of the most watched races in France - the Delage Type X racers came in first, third and fourth in a very close competition. The excitement and interest this created with the public convinced Delage to build bigger, more powerful models and to drop his one-cylinder cars.

After the 1913 races, the factory team cars were retired and new team cars were being built for the next season. W.F. Bradley, an American journalist and the European representative for the Indianapolis 500 Race, convinced Delage to sell him a car, and asked French champion René Thomas to drive it.

Pages 16-17 Racecar designed by Louis Delage and Augustin Legros. This voiturette is ready to race under No. 57 at the Coupe des Voiturettes in 1907, piloted by driver Pellegrin. It boasted a one-cylinder motor with a 100 x 120 mm bore and stroke and a radically lowered front profile.

Pages 18-19 1908 Delage racecar, Model G with a two-cylinder De Dion Bouton engine on a Type F chassis. It is currently exhibited at the Musée National de l'Automobile in Mulhouse with a large monoblock motor.

Above 1911 Type X racecar with three-liter engine designed by engineer Michelat. The team finished first, third and fourth in the Grand Prix de Boulogne on 25 June 1911.

Above right 1911, four-cylinder, three-liter Delage racecar, Model X. It is believed that this car raced at the Coupe de l'Auto in 1911. At the wheel is collector H. M. Petiet.

Below right Delage Type Y designed by engineer Michelat in 1913. It had a four-cylinder, 6.2-liter motor. The Y Model team placed fourth and fifth in the Grand Prix de l'ACF in 1913; first, second and fifth at the Grand Prix of France at Le Mans; and first and third at Indianapolis in 1914.

16
DUNLOP
CORD

Above Profile view of the Model D-II, six-cylinder, five-liter racecar. One of two cars made for hill climbs and sprints. Its first appearance was reported at Mt. Ventoux. Registration number is MV 2734.

Left Model D-II, 1922, Registration MV 2734 at speed.

Delage also consented to sell a second car to Delage dealer and driver Albert Guyot, provided that - if they won - the drivers could keep all the prize money, a very sizable sum. René Thomas won the race, collected his award, and was given the car to sell in the United States. He returned to France a rich and famous man. The Delage oval also became well known in the U.S. at no cost to Louis Delage.

Soon after this American racing adventure, World War II broke out and European racing was suspended. Delage lost his great chief engineer Michelat to the war effort and hired his cousin Charles Planchon, a gifted engineer. Once the war ended, the company took a new approach to racing and in the early 1920s developed a new five-liter car for sprints and hill climbs.

The car came in first at Mt. Ventoux, France, in 1922. After this and other successes in record car events, Planchon was charged with creating an engine to comply with the new rule changes imposed by the Automobile Club of France. He developed what would become his finest work: the 2LCV, a two-liter, dual overhead cam, V-12 cylinder motor, rumored to have been built in only 90 days. The car first appeared at the Grand Prix de l'ACF in 1923 but arrived late and did not have time to practice. It was not properly sorted out and broke down in the sixth lap. Delage was furious and fired Planchon, instructing engineer Albert Lory to make the engine work. This he quickly did. The car had a good record of successes through the 1924 and 1925 seasons, winning several races and achieving many podium finishes.

Above top This Delage 2LCV racecar has a very early V-12 motor with a twin camshaft. The main bearings are roller bearings, a revolutionary design.

Above Delage 2LCV racecar, V-12, two-liter Grand Prix, built for the 1923-1925 season.

Above right Delage 2LCV racecar No. 9 shown on August 1, 1924 after being weighed at the Grand Prix of Europe at Lyon. Driver Albert Divo, pilot for the official team, is shown on left of the photo, with co-pilot Fretet alongside. Two days later, the car came in second in the race.

Below right Montlhéry, Grand Prix de l'ACF, 1925. Driver Wagner is at the wheel of a 2LCV Delage.

ROLLAND
AUTOMOBILINE
bonne essence
fait
LES 24

The ACF changed the rules again for the 1926 season, which required a new motor with a maximum capacity of 1.5-liters. Anticipating this, Lory had already begun development of a straight-eight motor in 1924. The final product was a legend in racing history: a straight-eight, 1500 cc, supercharged, dual overhead cam motor renowned for its mechanical sophistication and reliability. The car's record for the 1926 season was good: third place at the Grand Prix of Europe and third place at the Grand Prix of Spain. The team came in first and third place at the British Grand Prix, and first at the Grand Prix de la Baule in France.

The 1927 season was even better, finally bringing Delage what he had hoped for: the World Construction Championship, with Robert Benoist driving all the winning races. The team came in first in the Grand Prix de l'Ouverture, first in the Grand Prix de l'ACF, first in the Grand Prix of Spain, first in the Grand Prix of Europe and first, second and third in the British Grand Prix.

At the end of the championship season, Delage had achieved his goal, but the effort had nearly exhausted the funds of the company forcing him to concentrate on beautiful coachwork on luxury chassis. The assets of the racing team were sold to private entrants who continued to campaign the cars. In 1929, Louis Chiron took one of the cars back to the Indianapolis 500 and finished in seventh place. This car still exists in its original condition at the Miles Collier Collection in Florida.

Above 1926 Delage 1500 created by engineer Albert Lory. In 1927 it gained Louis Delage the title of World Construction Champion, having won all the Grand Prix of the season. In this period photograph, Albert Divo is at the wheel.

Above left The engine of the car shown above, in its current condition. It is a Delage 1500 racecar created for the 1926 season. The engine capacity is 1.5-liters.

Below left Profile of the car above, shown restored. Its top speed was 210 km/hr.

Left 1926 Delage Model 1500 racecar, eight cylinders, two compressors. This car, also shown on pages 26 - 27, is part of the Miles Collier Collection.

The ERA Delage. In 1936, 27-year-old Prince Chula Chakrabongse of Thailand headed up a motor racing team called White Mouse Racing. From Dick Seaman he bought a 1500 Delage racecar for his cousin, Prince Bira, to drive. Uneasy about his royal neck inside a ten-year-old chassis, Prince Chula kept the drive train of the 1500 and commissioned Albert Lory to build a new 1500 racing chassis, into which they placed the old drive train. The White Mouse Garage then added a new body from drawings obtained from Louis Delage. The new chassis was very similar to the design of the original 1500 chassis, but improved with an independent front suspension, transverse semi-elliptical springs and Lockheed hydraulic brakes. In 1950, famous driver and team owner Rob Walker bought the car and blew up the Delage engine. No engine blocks were available from the Delage factory, which had gone out of business, so he went instead to the ERA Company and installed one of their motors on the chassis. This is a two-stage, supercharged E-type ERA engine. Peter W. Mullin Collection.

Above left 1939 Delage D6 racecar, Chassis 51820, three-liter, six-cylinder motor, driven at the Le Mans racetrack that year.

After 1927, the Delage Company stopped producing racing chassis entirely, but went on building chassis for road cars only. The surviving racing chassis continued to compete in international racing until the 1950s as private entries, with no financial support from the factory. In 1936, the success of these private entries led to the creation of two racing chassis for the Delage 1500, the World Championship model. Of these two chassis, one (featured on pages 30 - 31) became the ERA Delage.

In 1935, Louis Delage lost ownership of his company, which eventually fell into the hands of the Delahaye Company. Under Delahaye, production of racing chassis started up once more in 1936 when Walter Watney, owner of Écurie Watney, ordered a three-liter, six-cylinder Delage engine, placed on a modified Delahaye 135-S racing chassis, becoming Chassis 50688. The Delage motor for the car was a modified D6-70 engine that would become the engine for their post-war passenger cars. Watney commissioned Joseph Figoni to design a racing coupé on the chassis, which was entered to race at Le Mans in 1936. However, the race was cancelled, and instead it raced at Le Mans in the following year, coming to a fourth place finish. The Figoni body was removed and replaced with an open two-seat body for the 1938 race. In 1939, two new lightweight chassis were created (51820 and 51821). At Le Mans, one car came in second over all and the other retired early.

When World War II started, racing was suspended in occupied France. After the war, Delahaye built five D6 racecars with open two-seat bodies in 1946 for the 1947 season. They were campaigned successfully throughout Europe from 1947 to 1950. The horsepower had been improved from the pre-war D6 levels to 142 bhp at 5000 rpm. The weight was 950 kg. and the maximum speed was 210 kph. Chassis 880004, one of the five and the most campaigned of the group, was bought by Henry Louveau who drove the car in its first race to a second place finish at the Grand Prix of Perpignan in 1947. During that summer, he entered the car in the Grand Prix of Marseilles, where it achieved a third place finish. He then entered the car at the Grand Prix of Strasbourg and the Grand Prix of Comminges. Louveau did not always drive the car himself. It was Varet behind the wheel at the Grand Prix of Italy when the car came sixth.

The car in 1948 came in fifth at the Grand Prix des Nations and second at the Grand Prix des Frontiers. It ran in two more races, but did not place. In 1949 the car raced at Le Mans, with Louveau driving, and finished in second place. He also drove it at the Grand Prix of Spa, again coming in second.

Louveau entered the car at Le Mans in 1950, and came in seventh place. A short time later, the car was put up for sale and finally in 1953 it was bought by Jacques Nollé who had it rebodied in 1955 by coachbuilders Pichon & Parat as a coupé. Nollé had a hard time registering the car for racing, and sold it to Richard Gallagher, an American serviceman stationed in France. When Gallagher returned to the United States, he brought the car with him. It stayed on the East Coast for many years in its original 1955 condition, until 2002 when Peter W. Mullin purchased the car and restored it to its original 1947 racing configuration.

After the initial production of these five chassis, Delahaye dropped the Delage racing program. It marked the end of the illustrious racing history of the Delage Company.

Above Henri Louveau races a Delage D6, Chassis 880004 at the Grand Prix de l'ACF at the Comminges racetrack in 1949. Louveau, after a grueling race of more than 500 kms. and in torrid heat, placed fifth overall and third in his class.

Pages 34-35 Chassis 880004. 1946 Delage D6 with a factory body, Chassis 880004. Seen here at the Concorso d'Eleganza Villa d'Este, Italy in 2005. Peter W. Mullin Collection.

7

The Delage V-12 Labourdette

Louis Delage had long wanted to produce a closed race car as a prototype for a production car and finally managed to do so in 1937, when the French Grand Prix organizers changed the rules to allow closed cars to race. At the end of his career, when he was the head of S.T.A.S., his new small company, he created the Delage V-12 Labourdette, which can lay claim to several unique features: it was the only Delage designed entirely by Jean Andreau; the wooden model he built of the car was one of the first to undergo wind tunnel testing; and it was the first closed car designed to race at the French Grand Prix.

The production background of the car deserves our attention, as many hands took part in the project: Carrosserie Labourdette; the designer Jean Andreau; engineer Albert Lory; and three car manufacturers: Delage, Delahaye and the Société Technique Automotive Sport (S.T.A.S.).

Louis Delage set to work collaborating with expert Jean Andreau on the aerodynamic design, and with his former engineer at Delage, Albert Lory, on the development of a V-12 engine. It is interesting that Lory built his engine when the Delahaye V-12 engine was already in existence, in fact competing against it.

Jean Henri Labourdette, who had just created the Vutotal window system, was selected as the coachbuilder. The founder of Carrosserie Labourdette was Jean-Baptiste Labourdette, a blacksmith who traveled from the French Basque country to Paris and in 1858 started building bodies for horse drawn carriages. The firm built its first car body on a Panhard-Levassor in 1899. Jean Henri, grandson of the founder, started work at the shop in 1907 and his engineering and design skills quickly took him to the top. He was strongly influenced by the science of aerodynamics, and styled and produced a clean skiff design on a 1912 Panhard-Levassor. It was the first of many significant bodies that impressed his racing and aristocratic customers. Jean Henri Labourdette was a friend of brothers Georges and Gabriel Voisin and other aircraft manufacturers and these associations put him at the leading edge of streamlined styling. In 1934 he came out with an aerodynamic body for a Renault using a new pillarless technique on a four-door saloon. The concept was carried further in 1935 with a frameless windscreen and front door window that eliminated the "A" pillar by using structural glass for the windshield. The style was invented by Joseph Vigroux, patented as "Vutotal" and licensed to Labourdette. Its most exciting example is the 1937 Delage V-12 we discuss here.

Jean Andreau tested his design with a wooden model in a wind tunnel, and paid careful attention to the undercarriage of the body to make sure the car had minimum wind resistance. The wooden scale model is still in existence today. The tail fin was Andreau's solution to the problem of high-speed stability, and although a startling addition to the car's design it is one of Andreau's more conservative efforts. The Delage V-12, 4 1/2-liter prototype engine was placed in a Delahaye Type 135 chassis equipped with a special Cotal transmission. Finally, the car was ready to compete at the French Grand Prix in July 1937 at Montlhéry and participate in speed and endurance record-setting races. To the dismay of its creators, however, the car (Number 4) was damaged during racing practice by its driver, Mr. Fretet.

The car was repaired and appeared next at the Paris Auto Salon in October 1937 on the Labourdette stand. After the show, the coachwork was removed and an open two seat racing body was installed. The story of Jean Andreau's closed aerodynamic coupé therefore ends here. Its influence, however, is clearly visible in the designs of the Dubonnet Hispano and other models that came later.

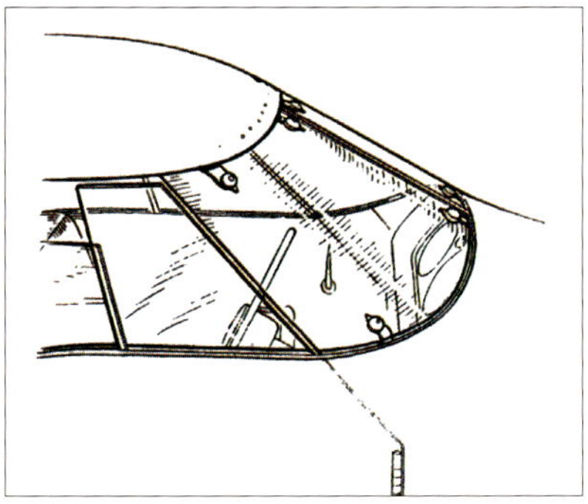

Above Detail of the Labourdette Vutotal structural glass windshield. An article in the journal of the English Society of Automobile Engineers describes it as follows: "The screen is a curved glass of unusual thickness, about 12 mm, and the same thickness is also used for the fixed quarter windows, since these, together with the screen, support the roof. These glasses are fixed into position by bolts passing through rubber-bushed holes in them. As the bolts are not visible in this design, the effect is extremely neat, and there is, of course, no obstruction to vision whatsoever. The curved screen glass is also reasonably free from distortion. Weather-proofness is assured by the curved screen glass overlapping the falling glasses of the doors, which, in turn, overlap the fixed quarter glasses."

Above right Profile of the Labourdette racecar designed by Jean Andreau and wind-tunnel tested. The car was Number 4 prior to the race, but never ran since it was damaged in practice.

Below right Delage V-12 Labourdette in practice before the 1937 Grand Prix de l'ACF at Montlhéry. The driver, Fretet, is on the left. The engineer, Lory, is on the right.

4

1000-RLI

DELAGE
DANDY

Chapter III

La Belle Voiture Française

With the triumph of the European Championship of 1927 behind him, Louis Delage was forced to face economic reality. It meant abandoning his racing dreams and turning the energy and focus of his company towards its passenger cars. At that time, before the invention of the monocoque body where chassis and body became a single unit, a chassis was manufactured separately, which the customer chose according to taste. Some customers liked the soft feel of a town car, while others preferred a faster, sportier chassis. Delage chassis were known for neat layouts and advanced technical innovations and had a reputation for speed and daring, carried over from the marque's triumphs on the racetrack. After a chassis was selected, the customer decided upon the coachbuilder, and a chassis with the prestige and quality of a Delage would naturally fall into the hands of the best of the best. It arrived "bare" at the coachbuilder's shop ready to be dressed, and customer and designer together defined what the body of the car would ultimately be: a closed coupé, a dashing cabriolet or, in some instances, a race car. The customer's particular wishes were considered, and lines and colors were carefully coordinated with haute couture gowns to create an elegant presentation at concours d'élégance and society events.

Delage started off with a new model for the 1927 Paris Auto Salon. It was the very successful DR, with a six-cylinder motor which sold 5,400 units over the next two years. Next year, his work force had reached nearly 2,000 people, and he decided that an eight-cylinder motor for a passenger car would be the proper evolution of his six-cylinder line. This time it would allow him to compete head to head with Hispano-Suiza and Rolls-Royce. Sales were looking better, and he appointed Augustin Legros, still his colleague in spite of many disagreements, to take charge of production, while Max Fallery supervised the marketing effort from his elegant offices on the Champs-Elysées. He asked Maurice Gaultier, who had created the six-cylinder motor that formed the basis for the entire Delage line, to start work on the new concept. It was to come as close as possible to the Hispano's powerful engine, sufficient to power the two-ton weight of his future model, but still respect the signature Delage style. Gaultier succeeded in producing an entirely classic motor, simple and solid in design. This was to be the Delage D8.

The D8 model was introduced in late 1929, moving the Delage Company away from the racetrack to brilliant performances at many concours d'élégance. It was a straight-eight motor that became an instant success in the high end market, a car of great luxury accompanied by top performance, uniting the best features of the race car and the town model. The chassis was well built, the engine silent and powerful and it was said at the time, "If it goes by swiftly and silently, it must be a Delage." The wheelbase was ideal - the right height, the right width, with beautiful proportions. The company had its own respected body department, but the chassis attracted the premier coachbuilders of the time, who applied all their skills to dress it in their most elegant and original creations. Letourneur & Marchand, Pourtout, Chapron, Labourdette, de Villars, Guilloré, Fernandez & Darrin, Franay and Figoni & Falaschi created some of their finest work on this chassis, producing what today have become automotive sculptures. These were the cars that dominated the concours d'élégance in the 1930s and allowed Louis Delage to rub elbows with the rich and famous. France gave a Delage D8 to Emperor Haile Selassie, King of Kings of Ethiopia; King Gustav V of Sweden was driven about in a Delage; singer Josephine Baker owned and adored several Delages; and although King Alexander

of Yugoslavia and French minister Louis Berthou were assassinated in Marseilles in October 1934 while driving in a D6, many politicians, movie stars, singers and royalty all over the world were delighted to be seen around town in a Delage.

The Delage D8 with its gorgeous custom-built coachwork entered the market in an era of crisis, a status symbol out of step with the times. The stock market in the United States and Europe crashed in 1929, and the Great Depression soon gripped the world. Delage missed the warning signs and continued to build new models, among them the fabulous D8-S, of which less than a hundred were sold. A six-cylinder model was then introduced in 1930 at a lower cost, along with other D8 models, but the company continued to struggle.

In 1932 all their models were being offered at prices people could no longer afford. Louis Delage was forced to start selling assets in 1933 and simultaneously made a last attempt to introduce new models to prompt sales. These did not materialize and Delage approached Peugeot and offered to sell his company. Still smarting from his defection in 1905 and anxious to eliminate a competitor, Peugeot expressed no interest and other rescue deals failed. Delage started to lose control of his enterprise to the banks, and on November 5, 1933 the creditors closed in and formed a corporation to liquidate the company. Temporary rescue came in the shape of Arthur-Leon Michelat, who was brought in to turn things around, introducing the D6-65 and the D8-105. The creditors waited and the new models were well received by the public and the dealers - but it was too late.

Pages 38-39 Henri Chapron gouache of Delage D6-75N model Dandy. Study of a cabriolet on a standard chassis.

Above 1931 Delage D8 cabriolet with a folding windscreen shown at a concours d'élégance where it won a Grand Prix d'Honneur.

Pages 42-43 1932 Delage D8 cabriolet with French President Edouard Daladier standing at the back during a tour of North Africa. Strangely, the car was driven with bare tires.

VIV
VIVE LA
LES CORS

DALADIER
FRANÇAISE
DE L'AFRIQUE DU

Capital was insufficient. Louis Delage tried to create another company to fund the new momentum but failed and the Delage Company was liquidated in 1935.

The Company's history now becomes complex, with several name changes and business arrangements with various investors. When the company was sold, Louis Delage went on to head a small firm called the Société Technique Automotive Sport (S.T.A.S.). At the same time, he was President of the S.N.A.D. (Société Nouvelle des Automobiles Delage), an enterprise charged with the marketing of the Delage marque. The buyer of the liquidated Delage Company was Walter Watney, the Delage agent for the Paris Region, and a great friend. Watney had purchased the Delage name at one point, but sold all his Delage assets to Delahaye, which then became the sole proprietor. At some moment in the ownership transaction, Watney and Delahaye licensed Louis Delage and his new company, S.T.A.S., to produce no more than 20 competition sport cars a year.

Delahaye continued to produce the Delage marque, and in 1937 launched the Delage D8-120, one of their finest automobiles. Cars still rolled out under the famous Delage blue oval until 1953, when Delahaye was forced to close its doors.

Louis Delage died at his chateau in Le Pecq in 1947, where he is now buried. In the words of his great-grandson Patrick Delage, "He was a fascinating personality, worthy of a novel. In his latter years, he retreated into a spiritual life, very far from the image that we normally have of him. This was under the leadership of Father Brothier, founder of the orphanage called Children of Auteuil."

La Belle Voiture Française ... these words that Louis Delage took to heart early in his career still stand for elegance and quality. His "Beautiful French Car" is now a cherished item in many of the most important car collections in the world.

Above top 1932 Delage D8-SS-100 bodied by Fernandez & Darrin. The three-position top with landau bars gives the car an elegant look. The chrome wheel covers were an option and well-chosen for this style.

Above Sales pieces for Autex, the company owned by Walter Watney.

Renée Friederich

Ernest Friederich had a Bugatti agency in Nice and was also a race driver for the marque. His daughter Renée was involved in women's racing at an early age, driving a Bugatti which Ettore Bugatti would lend her father. In 1932, Bugatti had a falling out with Ernest Friederich and turned down her request for a car. Louis Delage then stepped in and gave her the beautiful open D8-S car pictured on this page, which she planned to drive at the annual Paris to San Raphael Feminine Rally. On the second day of the rally, there was a hill climb which was a race against the clock. Renée was used to the much smaller Bugatti, and missed the turn coming out of an "S" curve. The car, shown here as it is today, rolled over on to its side on the soft shoulder of the road and although her mechanic was thrown out unhurt, Renée died of her injuries.

Above top Designed by Louis Boissier, chief designer for the Delage factory, this car was driven by Renée Friederich during her fatal accident. Only one of these cars was produced. Driver here is André Surmain.

Above 1931 Delage D8-S cabriolet bodied by the coachwork department at the Delage Company. The body has often been mistakenly attributed to Letourneur & Marchand. The car suffered only minor damage during the accident.

DELAGE
SELE

HENRI
CHAPRON
CARROSSIER
114-116 RUE ARISTIDE BRIAND
LEVALLOIS-PARIS

Henri Chapron

Henri Chapron opened his coach building workshop in 1919 in Neuilly-sur-Seine, near Paris. He had just completed his military service followed by an apprenticeship at a harness maker and carriage builder. New chassis were hard to find after World War I, so his first commissions were on military surplus Model T Fords. Chapron's first design was a torpedo style popular with the public, which brought him financial success and allowed him to expand into a larger facility at Levallois-Perret.

He was a prolific worker, ending up in the 1930s by creating top-notch custom built coachwork for leading brands such as Delage, Talbot-Lago, Hispano-Suiza and Ballot. At the beginning of the decade, the company had over 350 employees and produced more than 500 bodies a year, making it one of the largest producers in the country. When the Delahaye Company introduced its Model 135 at the Paris Auto Salon in 1936, its Chapron coachwork caused a big stir, and proceeded to win trophies at several concours d'élégance. Chapron then provided the coachwork for the D8-120, another of Delage's successful chassis.

Pages 46-47 Design study by Henri Chapron of a Delage D6 limousine for a three-liter car on a long chassis.

Above Early streamlined styling study by Henri Chapron for Delage model Mouette (seagull).

Above right 1938 Delage D6-70 bodied by Henri Chapron.

Below right 1939 Delage Grande Berline on a D8-100 long chassis bodied by Henri Chapron.

3819

During World War II, Chapron left Paris to run a small repair facility in the countryside, but returned after the war to create stunning bodies for Delahaye, Delage and then Rolls-Royce. When the market for luxury custom-built bodies collapsed in France in the early 1950s, it put the small independent craftsmen out of business and mass production became the order of the day. Henri Chapron started to work for Citroën in 1955 and continued through the seventies, ending with the Citroën-Maserati and a few custom-bodied Talbot-Lagos for General Charles de Gaulle and the King of Morocco.

Delage D8-120, Chassis 51760, order number 6089, delivered by Chapron on June 21, 1939. The price of the coachwork alone was 43,500 French francs. The chassis cost 105,800 French francs. Chapron was always careful to install his coachbuilder badge, showing the corresponding order number, on every car he built. It was generally placed above the rocker panel in front of the rear fender. The front and rear three-quarter images opposite show Chapron's typical elegant and conservative styling, in a two-tone color scheme. The car has been restored to its original condition. Peter W. Mullin Collection.

Chassis 51760 has a 4302 cc, straight-eight engine, with an 80 mm bore and 107 mm stroke. It has aluminum pistons with four rings and steel connecting rods. The crankshaft has four main bearings and the overhead camshaft has five main bearings. There is one two-barrel downdraft carburetor.

The car shown above was featured in the film *An American in Paris*, a 1951 MGM classic movie filmed on location. When Gene Kelly is not dazzling us with his dancing, he is seen being chauffeured around the city in the Delage. Painted green for the film, it was sent to Los Angeles and one of the members of the production team at MGM bought it and kept it for many years. Jim Hull and Peter Mullin acquired the car in 1987 and it is currently in the Peter W. Mullin Collection.

The designs of Henri Chapron were characterized by beautiful proportions and an elegant touch, and his work is considered a classic example of French style. He was also one of the few coachbuilders who managed to keep working after the end of World War II. When he died in 1978, his wife continued to operate the shop as a repair facility until 1986.

Pages 54-55 Chassis 51760. The car was designated by Chapron as a four-place cabriolet with a three-position top. It was judged at 100 points by the Classic Car Club of America. Peter W. Mullin Collection.

Left The instruments are exceptionally well-balanced on the dashboard and surrounded by wood trim. To the left of the steering wheel is the shifter for the Cotal electric planetary transmission.

Above The rear of this car has a "bustle back" with a covered rear spare and top loading trunk.

DELAGE
D8.120

Styling sketch of a 1938 Delage D8-120 bodied by Marcel Pourtout using a design by Georges Paulin.

Marcel Pourtout

Carrosserie Pourtout is an interesting story in the illustrious record of French automobile design. It is also an account of the collaboration among three men: Marcel Pourtout, the master coachbuilder; Emile Darl'mat, the Paris automobile dealer; and Georges Paulin, the inspired designer of aerodynamic bodies.

The Pourtout enterprise was launched in 1925 with three employees in a small shop in Bougival, near Paris. During their first years, they worked on nearly 500 cars, many of them Peugeots. Marcel Pourtout was close friends with Emile Darl'mat, whose thriving car dealership represented the marque. Together they collaborated on the production of a Pourtout-bodied Peugeot that they presented at the Paris Auto Salon in 1928. The success of the Carrosserie turned into commissions from prestigious customers, and Pourtout received chassis from the best manufacturers in the world: Bugatti, Voisin, Ballot, Minerva, Hispano-Suiza, Delahaye, Delaunay-Belleville, Talbot-Lago and Buick all carried Pourtout bodies. For Delage alone, Pourtout bodied close to 100 chassis among them several prize winning cars at the 1937, 1938 and 1939 Paris Auto Salons.

In 1933, Pourtout met Georges Paulin, who preferred working on advanced aerodynamic designs to his profession as a dental technician, and who thus joined the team. The three combined their talents to produce the revolutionary Peugeot 601, featuring Paulin's "Eclipse" design for a retractable steel roof. They went on to create the stylish Darl'mat Special Sport on the Peugeot 302 chassis.

Above 1933 Delage D8-S bodied by Marcel Pourtout. It has a fold down windshield with a three-position top. The hood ornament is by Lalique.

Above right Letter from the Delahaye Company to coachbuilder Marcel Pourtout concerning the 1938 Paris Salon show car, with English translation.

Below right Letter from the Delage Company to coachbuilder Marcel Pourtout with English translation.

DELAHAYE AUTOMOBILE COMPANY

Paris, March 9, 1939

Carrosserie Pourtout
39 Avenue Paul Doumer
Rueil-Malmaison

Gentlemen:

Please be advised that we have sold the Sports Coach built on a D8-120 chassis, number 51612, which was exhibited on your stand at the Salon.

We are returning the automobile to you to repair the metal fittings on the front seats, which were wrongly affixed to the base, and to modify the central arm rest at the back which is too high. In both cases, our inspector will provide you with all necessary information.

On another subject, our client has asked for the steering wheel to be raised to the maximum and we would be grateful if you would take care of this and adjust the dashboard for the passage of the steering tube.

It will also be necessary to repaint the wheels the same color as the leather, to install an ashtray on the dashboard, add monogram B.J.D. in block lettering, in the same color and style as the trim.

Please handle this as quickly as possible, since our client needs to take delivery of his car in the next few days.

Sincerely yours,
Soc. des Automobiles Delahaye

SOCIETE DES AUTOMOBILES

DELAHAYE

Société Anonyme au Capital de 10.867.500 Frs

SIEGE SOCIAL
BUREAUX & USINES
10 Rue du Banquier PARIS XIII
Tel. GOBELINS 43.00
Adr. Télégr. MORANAIN 77 PARIS

REPARATION
PIÈCES DÉTACHÉES
17 Rue Lebrun, PARIS XIII
TÉLÉPHONE
GOBELINS 37-88 43-00
MAGASIN D'EXPOSITION
75 Champs Elysées

Référence à Rappeler Ga/SL Service D.
Pièces Jointes

Paris le 9 MARS 1939

CARROSSERIE POURTOUT
39 Avenue Paul Doumer
RUEIL-MALMAISON
(S & O)

Messieurs,

Nous vous informons que nous avons réalisé la vente du coach sport qui a été exposé sur votre stand au Salon sur châssis D8-120 et qui porte le numéro 51.612.

Nous allons vous retourner cette voiture pour réparer les ferrures de sièges AV qui sont mal fixées au plancher, pour modifier l'accoudoir central AR qui est beaucoup trop haut; pour ces deux cas notre Inspecteur vous donnera toutes les indications utiles.

D'un autre côté, notre client demandant de relever la direction au maximum, nous vous demandons de bien vouloir échancrer la planche de bord pour le passage du tube de direction.

Il faudra également repeindre les roues de la même teinte que le cuir, de prévoir un cendrier sur la planche de bord, de poser des monogrammes, lettres bâtons, de la même teinte que la garniture aux lettres B.J.D.

Nous vous demandons d'activer le plus possible ces travaux, notre client devant prendre livraison de sa voiture ces jours ci.

Nous vous prions d'agréer, Messieurs, nos salutations distinguées.

DELAGE AUTOMOBILE COMPANY

Paris, March 23, 1937

Sir,

Further to our various conversations and to your letter of March 22, we are ordering from you a sports saloon, according to the drawing upon which we agreed on Saturday for the price of 30.000 Francs with the reserves mentioned in your letter of the 22nd of this month.

I am leaving immediately and count on you upon my return to find the full scale drawing entirely finished. I have noted the lead time of two months which you have indicated and count on you to make all necessary efforts to keep to this schedule. Take great care in making this car, this can be as interesting for you as it is for us.

As soon as your full scale drawing is finished, please send me a few words care of the Toulon (Var) Post Office, and I will cut short the few days of vacation that I am taking so that this car can be ready as soon as possible.

Yours very truly,

L. Delâge

P.S. If during the preparation of the full scale drawing you encounter any stumbling blocks, please do not hesitate to call Mr. Baillemont, our engineer at Delahaye, who can come and meet with you. I have already advised him, and he is available to assist you immediately.

SOCIÉTÉ NOUVELLE
DES
AUTOMOBILES DELAGE

Société Anonyme au Capital de 1.000.000 de Francs
R.C. Seine 267.952 B

MAGASIN D'EXPOSITION
SERVICES COMMERCIAUX
23.25, Av. Victor-Emmanuel III, Paris VIII
Téleph. BALZAC 40-00
Adr. Télégr. DELAJAUTO. PARIS

PIÈCES DÉTACHÉES
RÉPARATIONS
42 Rue de Campo-Formio, Paris XIII
Téleph. GOBELINS 43.00
Adr. Télégr. MORANAIN 77 PARIS

Code Lieber
Code A.B.C. 5th et 6th
Code Bentley
Code A.Z.
Code The M.T. Telegram

LD/JP.

PARIS, le 23 Mars 1937.

Monsieur Marcel POURTOUT
Carrossier
39, Avenue Paul Doumer
RUEIL-MALMAISON
(S & O)

Monsieur,

Comme suite à nos divers entretiens et à votre lettre du 22 Mars, nous vous passons commande d'une Conduite Intérieure Sport, conforme au dessin que nous avons arrêté ensemble samedi, pour le prix de Fr 30.000.-- sous les réserves que mentionne votre lettre du 22 courant.

Je pars à l'instant, et je compte sur vous pour qu'à mon retour, je trouve le grand plan parfaitement au point.

Je note le délai de deux mois que vous nous indiquez, et je compte sur vous pour faire tous vos efforts pour le tenir.

Soignez bien l'exécution de cette voiture, celà peut être aussi intéressant pour vous que pour nous.

Dès que votre grand plan sera achevé, envoyez-moi un mot Poste Restante à TOULON (Var); j'écourterai, s'il le faut, les quelques jours de vacances que je prends, de façon à ce que cette voiture soit prête le plus vite possible.

Agréez, Monsieur, mes salutations distinguées.

Delâge

P.S.- Si en cours d'exécution de votre grand plan, vous étiez arrêté par quelque chose, n'hésitez pas à demander à BAILLEMONT, notre Ingénieur de chez DELAHAYE, de venir s'entretenir avec vous. Je l'ai prévenu, il se dérangera immédiatement.

Just before World War II, the Rolls-Royce representative in Paris, Walter Sleator, sold a Bentley chassis to Mr. André Embiricos, a Greek banker living in France. He recommended Pourtout for the coachwork, and Georges Paulin created a wooden scale model that was tested at the Vickers Aircraft Company. The aerodynamic results and the elegance of the design convinced Bentley to bless the project, which became the Embiricos Bentley, one of the most beautiful examples of aerodynamic design ever produced. Characteristically for Paulin, the lines of the car are allowed to speak for themselves, with no distracting chrome trim or other design elements.

When World War II broke out, Claude Pourtout, the founder's son, recalls that he and the remaining staff at the shop built the prototype of the American Volunteer Corps ambulance on the Chevrolet W158/2 chassis. Just before the Germans invaded, the ambulances and twenty-two of the wives and children of the workers fled Paris. As with many members of the coachbuilding world, Marcel Pourtout was a member of the French Resistance, but remained undiscovered to the end. In fact, the Germans conscripted him into being the Mayor of Rueil-Malmaison, where he kept up his dangerous undercover activities. Carrosserie Pourtout returned to business after hostilities ended in 1946, but times were tough. There were few rich customers, and even fewer luxury chassis. Finally, in 1948 Marcel Pourtout handed control of the company to his son Claude, who went on to build the passenger deck for the Paris tour boat "Bateau Mouche" and also construct industrial advertising vehicles.

Marcel Pourtout served a second term as Mayor of Rueil-Malmaison and this became his principal activity. He died in 1979, aged 85.

Pages 62-63 Delage D8-120, Chassis 51612, the 1938 Paris Show car.

Above 1938 Delage D8-120, Chassis 51612. The model designation is a coupé profilé. It was built by Marcel Pourtout and designed by Georges Paulin, who considered it one of the sister cars to the Embiricos Bentley. The original price of the chassis was 105,800 French francs, and the Pourtout coachwork cost an additional 35,000 francs. It was built for the Paris Auto Salon of October 1938.

Above right The restored Chassis 51612. The styling is clean with soft tear drop shapes repeated in the pontoon fenders and tapered roof line.

Below right Chassis 51612. The rear shape of the car is inspired by the Talbot-Lago tear drop design.

Louis Delage's Own Car

Louis Delage commissioned a Delage D8-120-S, Chassis 51620, from Marcel Pourtout as a prototype show car for the 1937 Paris Auto Salon, and ended up by using it himself. Clearly Delage was very interested in the quality of this model, and exchanged several letters with Pourtout on the matter (see page 61). Built on the D8-120-S chassis, the suspension was lower than standard. The coachwork was all aluminum, with an engine bored out to 4750 cc, larger than any other, and giving the car more power. The combination of these features created a high-speed, aerodynamic coupé. The car is in the Samuel J. Mann Collection.

Above Chassis 51620. Georges Paulin balanced all aspects of the styling and produced a coordinated, streamlined masterpiece which today has the look of a contemporary prototype. The aerodynamic shape was tested at the wind tunnel facility in France, a concept then in its infancy.

Page 68 The body tapers from front to rear in a set of continuous curves. At a cost of 30,000 French francs for the coachwork, it was one of the most expensive cars produced by Marcel Pourtout.

Page 69 above The traditional upright Delage radiator leans back into the body to create a more graceful profile.

Page 69 below All the shapes meet at the tapered rear.

DELAGE

DELAGE

Left A styling sketch of the Delage shown above, featured in En Route magazine which contained a review of the 1937 Paris Auto Salon with particular praise for the car.

Above Delage D8-120-S, Chassis 51620, a personal show car for Louis Delage, designed by Georges Paulin. This Delage was built in 1938, at the same time as the Embiricos Bentley. It is in the Samuel J. Mann Collection.

Right The rear wheels were covered to eliminate wind resistance.

Carrosserie de Villars

Carrosserie de Villars was a small company in Courbevoie. The company had begun as a repair garage outside Nice, started by Francis J. Gould, an American millionaire who owned real estate in the area. Gould put together a body shop in 1925 for the purpose of restoring his friends' cars and in part to occupy the leisure time of Roland de Graffenried de Villars, his son-in-law. When the marriage of the de Villars ended in divorce, de Villars left the business, but his aristocratic name remained behind.

The stock market crash did not seem to have deterred Gould, who decided to expand the shop in 1929 to build the coachwork for luxury French cars. With a tough manager at the helm of a small group of craftsmen, the de Villars firm produced no more than twenty-five bodies a year. They were built to the highest standards, using only the best materials and strictest production techniques. So exacting were these standards that Gould would not allow a bead to be used between the fenders and the body. Cars were road tested before painting to ensure that not a single squeak or rattle could be heard from the bodywork. As a result, the shop commanded top clients, among them Prince Aly Khan, Prince Gourielli, husband of Helena Rubinstein, and the House of Monaco, which ordered six cars between 1930 and 1939.

The company did not survive the war. It was sold in 1945 to Jean Daninos and became a Facel Garage.

Above A styling sketch of a Delage D8-120 with coachwork by de Villars. The actual car, Chassis 51626, is in the Samuel J. Mann Collection.

Above right Profile of the Delage D8-120, Chassis 51626. Note the wire wheels.

Below right Rear view of Chassis 51626, showing the fin down the lid of the trunk.

Pages 74-75 Chassis 51626, front three-quarter view. Note the standing riveted seams on all four fenders.

delage

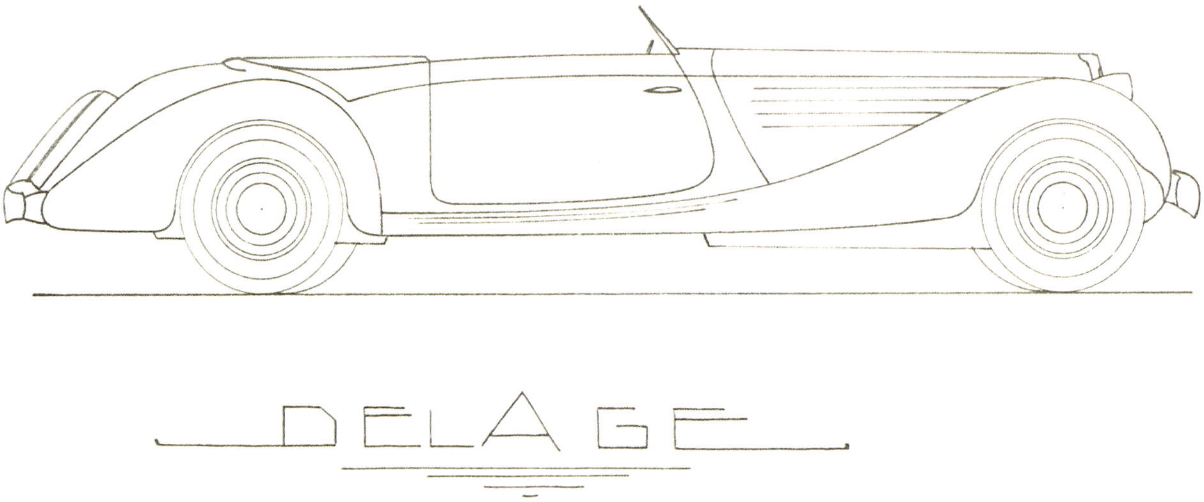
DELAGE

7675
ETABLISSEMENTS

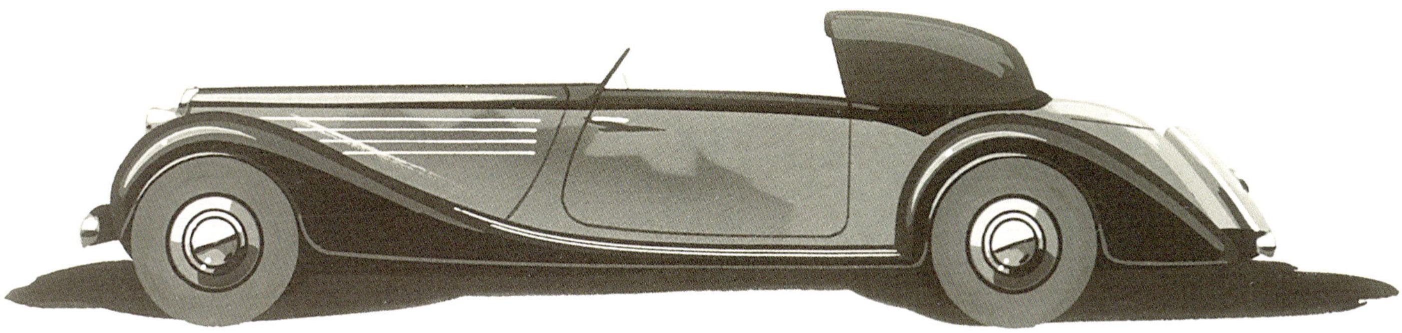

IGONI & FALASCHI

Figoni & Falaschi

The influence of the ideas of Figoni & Falaschi on the custom coachbuilders of the 1930s cannot be overstated. It is still felt by modern car designers today.

The company was started by Joseph Figoni, who was born in 1894 in Piacenza, Italy and brought to France when he was three. At 14, he was apprenticed to Émile Vachet, a charron, or wagon maker, one of those professions that evolved into the new business of automobile repairs and coachwork. He was called up for military service in 1914, spent seven years in the army, and then opened up a body repair shop.

Pages 76-77 Styling study of a Delage cabriolet bodied by Figoni & Falaschi.

Above top Styling study of a Delage D8-100 coupé bodied by Figoni & Falaschi. It is shown in two colors.

Above Styling study of a Delage D8-100 cabriolet by Figoni & Falaschi, with a three-position top.

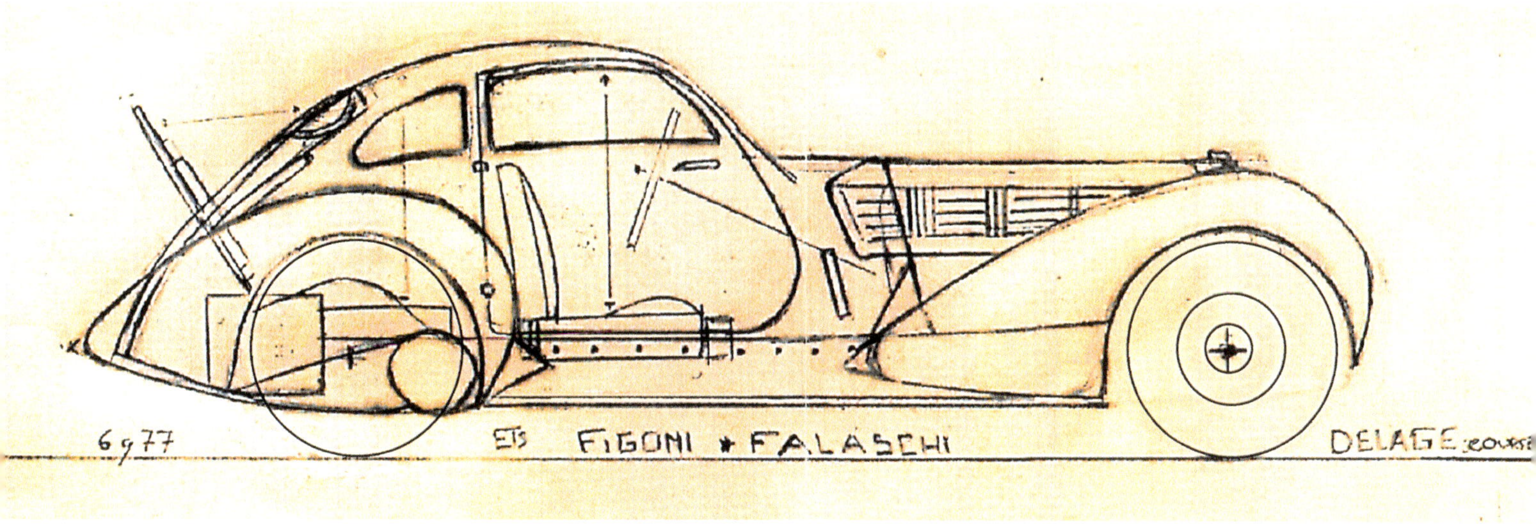

At first, Figoni worked with chassis from Panhard, Bignan, Salmson, Bugatti, Renault and Delahaye. By 1927 he was making bodies for Italian sports cars such as Lancia and Alfa Romeo. One of his racing triumphs was the aerodynamic design of the 1932 Alfa Romeo 8C 2300, which won the Le Mans 24-hour race in 1932 (driven by Raymond Sommer and Luigi Chinetti). In 1933 a Figoni-designed Alfa Romeo driven by Tazio Nuvolari won again and Chinetti came in second.

Above top 1936 Drawing by Figoni & Falaschi of a Delage D6-70 competition and concours car.

Above This Delage was built to race at Le Mans. The race was cancelled because of a strike in 1936 and the car was later presented at a concours d'élégance. It eventually went on to race at Le Mans in 1937.

Figoni's attention to detail and talent as a designer quickly won him business with customers who wanted to modify their touring cars, and from there he gradually moved on to full designs of custom coachwork. By 1925 he was building complete bodies on a variety of chassis and in 1929 placed his first body on a Delage. His design for the new Delahaye 18CV was introduced at the Paris Salon in 1933 and soon he was patenting his ideas. In 1935 he produced highly original bodies on Talbot-Lago and Delahaye chassis, including the Model 135 Delahaye, a sensation at the 1936 Paris Auto Salon. Following this success, Figoni designed an icon of the streamlining era, the Talbot-Lago T150-C-SS coupé.

Figoni's approach to his designs was totally "hands-on." He started with a simple sketch, and then reproduced the design in full scale with narrow iron straps attached to the chassis, creating a mock-up of the body. Work then moved from this maquette to the contours of the hand-formed sheet metal supported by wooden framing. Later, Figoni worked with modeling clay and refined the shapes while enlarging the guide templates to full size, a process that was better for depicting the designs inspired by the model of the Couzinet airplane he kept in his office, with its tear drop shapes. His partner Ovidio Falaschi explained: "We really were true couturiers of automotive coachwork, dressing and undressing a chassis one, two, three times and even more before arriving at the definitive line that we wanted to give to a specific chassis-coachwork ensemble."

Above Delage D6-70 bodied by Figoni & Falaschi. It was styled following the design for the New York World Fair car of 1939 on a smaller chassis. At the wheel is singer Tino Rossi.

Above right Rear view of a 1939 Delage D6 cabriolet, chassis number 51740, bodied by Figoni & Falaschi.

Below right Chassis 51740, by Figoni & Falaschi. It has three carburetors.

World War II was the beginning of the end for custom coachwork. The Figoni & Falaschi workshops were requisitioned by the French government to make aircraft components. When France fell to the German occupation, Figoni left for the Dordogne, and upon his return to Boulogne-sur-Seine he turned to the manufacture of household appliances. After the war, new industrial regulations and tax laws gradually doomed the French coachbuilding industry, and the company, once a world leader in the field of automotive design, started new ventures. General Motors tried to hire Figoni to come to the United States as a designer, but he was unwilling to leave his family and clients. In 1952 Figoni produced the Citroën 15-Six and received the *Prix de l'Art et de l'Industrie* at the Paris Salon. He then created two Delahaye 235s and sixteen Simca Sports, but no further custom coachwork ensued.

After 1954, the firm continued business as a Lancia dealer until 1989, under the direction of Claude Figoni. "Ironically," says his son Claude, "because of the demand for custom coachwork from the United States, we made more money after the war than before the war, which was unique for French coachbuilders."

One of the many tributes to follow in future years came in 1951, when the Museum of Modern Art in New York presented "Eight Great Automobiles," and an absolutely essential element of the exhibit was a Talbot-Lago coupé bodied by Figoni & Falaschi. Their designs are considered to be examples of modern art, and continue to win top prizes at all the major shows throughout the world.

Joseph Figoni retired to Boulogne, where he died in 1978 at the age of 84.

1939 Delage D6, three liters, bodied by Figoni & Falaschi, at the Paris Bagatelle.

Fernandez & Darrin

Howard "Dutch" Darrin and J. Fernandez formed a partnership to create coachwork in Boulogne-sur-Seine near Paris, from 1932 to 1937. Darrin, a legendary American designer, had just shut down his French coachbuilding operation Hibbard and Darrin in 1931, due to the stock market crash and ensuing Depression. In 1932, with capital provided by his new partner J. Fernandez, a South American banker living in Paris, he continued his early work on luxury cars of the day. Darrin was the chief designer and Fernandez handled the financing. They created original and exciting styling on Delage D8, Duesenberg, Packard and Mercedes-Benz chassis. The market for coachbuilding on luxury cars declined in 1936 and the company ceased to operate the following year. Darrin returned to the U.S. and settled in Hollywood, where he created a new styling studio he called "Darrin of Paris."

Left Chassis 35875. The two-passenger boat-tail styling for the D8 was unique. In France, this was known as "naval" styling and was very popular. The original windscreen folds down, but the wire wheels in the car shown have been updated.

Below 1933 Delage D8, Chassis 35875, bodied by Fernandez & Darrin. It was styled following the design of the Auburn boat-tail speedster.

Above right 1932 Delage D8-SS-100 cabriolet bodied by Fernandez & Darrin. It handles like a sports car, fast and fun to drive. The model helped to build the reputation of Delage among sports car aficionados and racers.

Below right 1933 Delage D8-100-S bodied by Fernandez & Darrin in Paris. The front and rear fenders connect in a sweeping curve. The car has a disappearing top and a special 4-carburetor set-up for the engine.

A. Guilloré

A. Guilloré started his firm in 1937 in Courbevoie-sur-Seine, the district near Paris where many of the coachbuilders had their shops. His pre-World War II production was very limited but eventually the shop expanded and became well known for its post-war work on Delahaye, Delage and Talbot-Lago. It was featured frequently in their factory advertising. Guilloré's styling often used chrome trim to create accents, and his hood design with small round circular vents became one of his signatures. Guilloré produced a very diverse line in luxury vehicles, ranging from berlines and limousines to coupés and cabriolets.

Below far left Remarkable 1938 Delage D6-75 cabriolet bodied by Guilloré. The somewhat rigid roof line contrasts with the softer styling of the body. The round air vents in the hood are signature Guilloré styling.

Above 1938 Delage D8-120-S, Chassis 51754, bodied by Guilloré. The car is provided with two spare wheels and Vue Panoramique windows.

Left The chrome exhaust pipes were a signature of the D8-120. The screen mesh was to help dissipate heat from the engine compartment.

20
988-RQ 3

After World War II, Guilloré was chosen by Delahaye as one of the company's favorite designers. The shop's coachwork was conservative with an envelope-type body, influenced by post-war U.S. designs. Bodies were produced mostly in steel, not aluminum, and tended as a result to be heavy but well constructed. Guilloré participated in many concours d'élégance, winning top prizes from Monaco to Paris.

Some of his most famous models were the "Bearn" on Delage D6 and Delahaye 135 chassis, the "Vendôme," a berline on a Delage chassis, and the "Deauville" and the "Esterel" cabriolet on a Delahaye 135.

The Guilloré shop also produced vans and buses for businesses to publicize their products through tours of the French provinces and Europe. His coachwork on cars ceased in 1950 but the specialty truck business continued.

Above left 1949 Delage D6, three liters, bodied by Guilloré, at a concours d'élégance.

Below left A beautiful Delage D6 cabriolet, bodied by Guilloré, at Bagatelle.

Above File photo, circa 1951, of Delage D6 styled by Guilloré. Note the slab sided styling of the body below the belt line, typical for its time.

Carrosserie Franay

Jean-Baptiste Franay began his career as a saddle maker. He eventually went to work as a foreman for Henri Binder, a well-known and well-regarded coachbuilder in Paris, but left in 1903 when he was forty-five years old to start his own shop in Levallois-Perret near Paris. In spite of his late start, his firm became a successful small producer of quality coachwork, which in its early years was renowned for its conservative style. Jean-Baptiste's son Marius came to work for his father after he finished school and took over the business upon his father's death in 1922. Marius had a more creative and elegant approach which he applied to the luxury chassis produced by Delage (D8-15, D8-100 and D6-70), Duesenberg, Packard and Hispano-Suiza.

When the Delahaye 135 model came out in 1936, Franay Coachbuilders formed a close tie with the company which lasted until Delahaye's production ceased in 1953. Franay is particularly noted for the streamlined lines of the 1946 Delahaye, a former race car, and for a wild 1955 Bentley over-embellished with chrome. As was the case with all the small independent luxury coachbuilders in France after World War II, business slowed down and one of the last bodies to be produced by the shop was on a Citroën 15-Six in 1955, the year the company came to an end.

Above left 1937 Delage D8-100 town car, with coachwork by Franay.

Above Side view of Franay's 1937 Delage D8-100 town car.

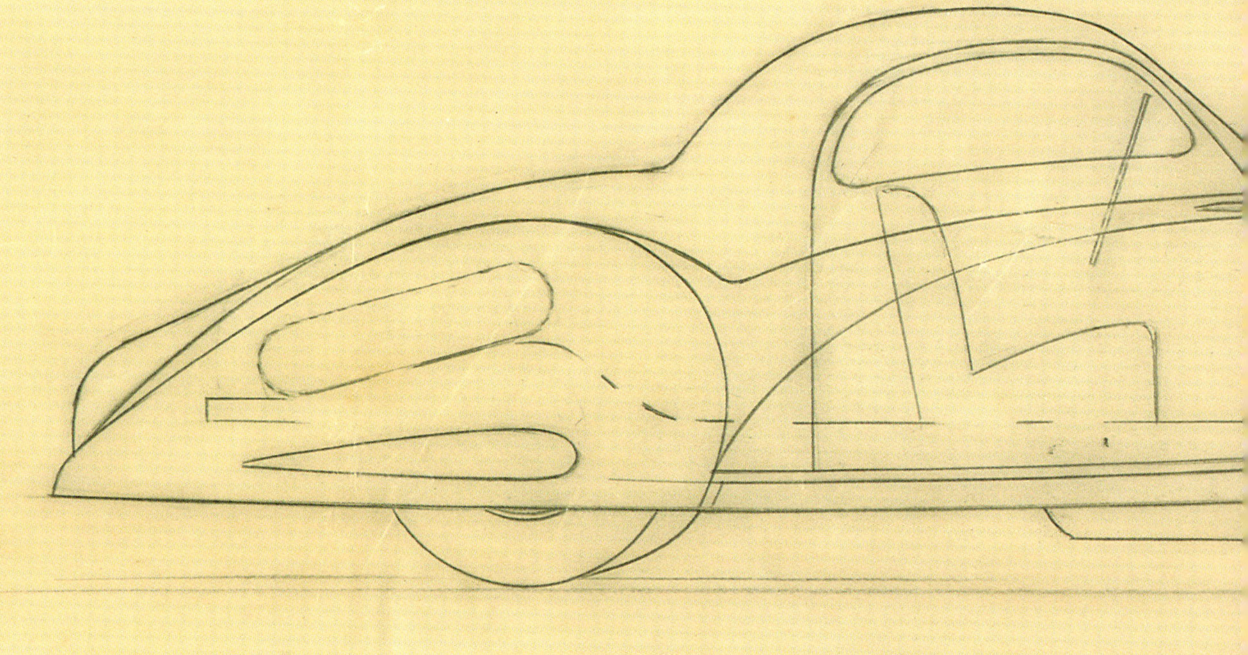

Dessin No 5719
D. 8. 120. Delage
Maison Delage

Letourneur & Marchand

Jean-Marie Letourneur and Jean-Arthur Marchand started their workshop in 1905 in Neuilly-sur-Seine on the outskirts of Paris. Both craftsmen met while working for Henri Binder, a leading coachbuilder in Paris who had started building automobile coachwork three years earlier, producing bodies for Delahaye and Panhard chassis. In addition to coach building, the two men had an interest in aviation, and built wings and fuselages for racing monoplanes before World War I. One of these planes was the first to fly from Paris to London. Letourneur & Marchand were capable businessmen, always ready to diversify into new technological areas, and as a result the companies founded by them were able to operate well into the 1990s.

The partners started working as sub-contractors for other coachbuilders and, during this time, also created a limited number of custom bodies. They were successful, and in 1910 they built new premises and expanded the business. From Gustave Baehr, an inventive Paris coachbuilder, they acquired a new patent for the all-weather "transformable," one of the first patents for a convertible top, and one which offered true weather protection and sound deadening. The technology was immediately incorporated into their new cabriolets.

Throughout the world wide depression of the 1930s, Letourneur & Marchand continued to receive major commissions from customers who loved the quality and beauty of their cars and whose fortunes were impervious to the economic downturn. Prince Nicolas of Romania ordered a Duesenberg and entertainer Josephine Baker appeared regularly at the various concours alongside her latest Delage. These clients chose chassis built by Bugatti, Voisin, Rolls-Royce, and Hispano, and brought them to the shop to be coached into luxury cars. Delage however, remained the firm's main customer.

Marcel Letourneur joined his father in the firm in 1928 after studying design in England. He brought in an Italian chief designer, Carlo Delaisse, and the company entered the streamlined decade. The team created some of the best styling of the era on the Delage D8-120 chassis, and these cars remain classics today. From 1931 through 1939, they built 115 custom ordered bodies on the Delage chassis as well as thousands more under contract to Delage and then Delahaye through their subsidiary, Autobineau.

Once the war was over, and after the deaths of Jean-Marie Letourneur and Jean-Arthur Marchand, designer Carlo Delaisse headed up the post-war styling for the company. Some of his featured designs were for the Delahaye 175 and Jaguar XK120. But the heyday was over: the era of custom-bodied chassis was ending and from 1947 to 1952 only sixty-seven bodies were built. The company finally ceased operations after completing a Renault contract in 1959. However, it is interesting to note that the Letourneur & Marchand company survived through its subsidiary, Autobineau, created to produce production cars for Delage and other marques. From 1925 to 1952 Autobineau went on to produce thousands of bodies on different models of Delage chassis. After 1952 and into the 1970s, Autobineau produced truck cabs for Unic. But at this point their involvement with the streamlined movement had long since come to an end.

Pages 92-93 Drawing of a D8-120, style 5719, by Letourneur & Marchand, for Monsieur Louis Delage.

Above right Letourneur & Marchand created the radical "Yo-Yo" car, so called because of the round louvers on the bonnet with interlocking tear drop shapes. It was a signature design for the Art Deco style, built especially for display at concours d'élégance. The car is a 1934 Delage D8-15 coupé with a sliding sun roof. It is presented in two colors, with rounded doors that repeat the motif of the hood. It is also an early example of fastback rear styling. Unfortunately, the car did not survive World War II.

Middle right The customer ordered the car starting from a concept drawing. This is a styling study of a Delage D8-120 cabriolet bodied by Letourneur & Marchand.

Below right A 1939 Delage D8-120 cabriolet bodied by Letourneur & Marchand created according to the concept drawing. It has a luggage compartment and a front bench seat that will at a pinch accommodate three people.

Above left 1936 Delage D8-120, Chassis 51012, styled by Letourneur & Marchand. It is designated by the coachbuilder as the model Aérosport. Eleven of these cars were built over a three-year period, in two series. The first series consisted of six cars.

Below left Side view of Chassis 51012. It is the first in the initial series of six cars. In October 1936 it was displayed at the Paris Auto Salon, together with other models created for the 1937 season. Here it is shown at the Meadow Brook Concours d'Elegance. James Patterson Collection.

Above Chassis 51012 in its current restored condition. The early cars had metal louvers on the hood as opposed to the wire screen that came later. James Patterson Collection.

Left Chassis 51763. In line eight-cylinder, four-liter engine with single carburetor. The exterior exhaust was a signature of the D8-120 model. This car is also shown on the following page. Arturo Keller Pyramid Collection.

Above 1939 Delage D8-120 Aérosport, Chassis 51763, by Letourneur & Marchand. The model was chosen by the French Government to represent the French automobile industry at the 1939 New York World Fair held at Flushing Meadows, and created a lot of excitement with the American public. This is a rear three-quarter view. Arturo Keller Pyramid Collection.

Left 1938 Delage D8-120-S Aérosport, Chassis 51018, bodied by Letourneur & Marchand. The fin that travels from the top of the roofline to the tip of the tail makes this car unique. The car was ordered by Shah Wali Khan, Viceroy of Afghanistan. It was moved to France and survived World War II hidden in a wine cellar behind a false brick wall. Discovered in the 1990s, it was restored and is now exhibited at the Petersen Automotive Museum in Los Angeles.

Right 1939 Delage D8-120-S, Chassis 51755, bodied by Letourneur & Marchand, showing fine chrome trim. This car is unique on a short chassis. James Patterson Collection.

Above left 1939 Delage D6-75 cabriolet bodied by Letourneur & Marchand, with a 3-person bench seat and luggage compartment.

Left Delage D6-70 bodied by Letourneur & Marchand. It features a Vue Panoramique window, a unique Letourneur & Marchand patent that did not use a pillar. The cloth roof incorporates a rear window, which slides back on tracks until it is neatly folded and stored above the trunk, as shown in the photograph.

Driving Impressions

The Delage Journal, published in Great Britain by the Delage Section of the Vintage Sports Car Club, is a treasure trove of information for all admirers of the marque. In its first issue, its editors stated that their aim was "to combine items of historical interest with contemporary articles." Among these are two interesting pieces that are reprinted here with kind permission from the *Journal*.

The first, *The D8-120: Excellent Performance of an Extremely Comfortable Car,* was first printed in *Motor Sport* in May 1938, and *The Delage Journal* reprinted it in issue No. 3, Autumn 1968. It offers a vivid driving impression that is just as fresh today as it was on the first day of print.

The second, *The D8-120: A Road Test,* (No. 13, Autumn 1974) is a more technical approach that will appeal to anyone with an interest in diving under the hood of a Delage. "Not since our issue No. 3 in Autumn 1968," say the editors of the *Journal* in a brief introduction, "have we looked at this model, which is a somewhat *rara avis* in the United Kingdom. On that occasion we published a road test from *Motor Sport* for May 1938. Below, by way of interesting comparison, we reproduce the *Autocar* road test of the self-same car - MG 5320 - from the issue of April 15th, 1938".

It bears noting that in the 1930s the most valuable portion of a car was its chassis. It was expensive and not always easy to obtain, and its number is always tied to the car's identity. The body of the car, on the other hand, received less respectful treatment and was often removed or modified. A cabriolet body might be installed on an individual chassis for open air driving during the summer months, and then changed in winter for a coupé, protecting the driver from the weather.

The car described here (registered in Great Britain as MG 5320) arrived in England in 1937 as a showroom demonstrator for University Motors, the Delage dealership. It had its original "spicy" French shape, which only a "trained contortionist" could leave or enter. This description reflects British tastes of the day, somewhat disparaging of the Art Deco lines from across the Channel, and inclined towards open conservative styling. University Motors was eager to sell the car, so its closed aerodynamic body was speedily removed. The British coachbuilding firm Coachcraft Ltd. was contacted and their chief designer Geoffrey Durtnal was asked to create a new body more attuned to restrained English tastes. Durtnal, however, had just spent six months studying and working with Henri Chapron at his shop in France, and produced a four place cabriolet, or drophead foursome coupé, as this style was known in Great Britain, that is remarkably reminiscent of a Chapron car. A Chapron car with a design very similar to Durtnal's can be seen on pages 51 and 54-55 of this book. It is amusing, therefore, that when the writers of the articles praise the Coachcraft body they are in fact admiring a design inspired in France.

The D8 - 120:
Excellent Performance of an Extremely Comfortable Car

Monsieur Louis Delage has probably turned his long and distinguished racing experience to better account in his production motors than most people. Ever since the war the name of Delage has been synonymous with refined speed, and, indeed, the man who could design the 1 1/2-litre 1925 Grand Prix machine might be expected to make a touring or sports-car that would do its job without a lot of fuss and clatter.

TOURS MINUTE
x 100
JAEGER
HUILE
CODE
AVERT

Above A front view of the Delage D8-120, Chassis 51018, by Letourneur & Marchand. Note the extra tear drop ornamentation on the rear fender skirts. The car is in the Petersen Automotive Museum.

This characteristic has persisted throughout the years; whether in the just-post-war 24 h.p. six-cylinder that was ten years ahead of its time, the famous DISS and DMS series of the later '20s, or the present models.

The biggest of the latter is the D8-120 of which there are as yet only a few in this country. When the opportunity of a short run in one recently presented itself it was therefore thought that the readers of *Motor Sport* might like to hear something about its performance.

The particular car in question had had a somewhat chequered career. It was originally fitted with one of those particularly spicy streamlined saloons which the French seem to like, but which none but a trained contortionist can leave or enter. On arrival in England, where it was destined for use as one of the demonstration cars of University Motors Ltd., who are the sole concessionaires in England, and the British Isles, the body was therefore removed, and a drophead foursome coupé by Coachcraft Ltd. put in its place.

In this form it had traveled nearly 11,000 very hard miles when I drove it, during which time there had been little opportunity to attend to it, so that its performance may be taken as being probably not quite as good as might be expected in private ownership.

On the road, I have not come across a really fast car that can inspire confidence more quickly. After being driven for a few miles by Mr. John Bradstock, a Director of University Motors and one of the best drivers of my acquaintance, I took over and found myself exceeding 90 m.p.h. within the first mile without exciting any signs of marked agitation on the part of my passenger. The performance, indeed, is formidable, but I think the roadholding and steering are even more remarkable. At whatever speed, the car can be put right in the gutter without any shock reaching the passengers or the stability of the car being affected, and there is absolutely no rolling, even when it is forced round a traffic island with all four tyres squealing. Equally, on Faversham Margate Coastal road, the many fast bends were negotiated with complete stability and comfort between 80 and 90 m.p.h.

Above 1938 Delage D8-120 Aérosport, by Letourneur & Marchand. A Delage Company file photo.

Pages 106-107 1939 Delage D8-120 Aérosport, Chassis 51763 by Letourneur & Marchand. The side windows drop down below the waist of the car, with the chrome trim accentuating the line. Arturo Keller Pyramid Collection.

The steering is finger light at all speeds and yet requires only 2 1/4 turns of the steering wheel to move the road wheels through a really wide lock. That this should be possible with 7 in. section tyres is indeed an unflattering reflection on those manufacturers who cannot apparently produce light steering that requires less than 4 1/2 to 5 turns from lock to lock.

The man who pretends that it is possible to corner fast and safely with low geared steering is either a fool or a salesman.

Another characteristic of the Delage steering is that it is completely devoid of "feel", and how it has been possible to combine this with such extraordinary accuracy and high-geared operation is obscure to me. To anyone used to normal sports-car steering, especially of the more "Vintage" types, this seems very strange at first, but the feeling rapidly passes. To begin with, one is apt to "jaggle" with the steering wheel in an attempt to "feel" the front wheels, but the temptation should be sternly resisted, as the best results are achieved by a steady even movement of the wheel, when the car will go round bends as though on the traditional rails.

This lack of reaction in the steering is met elsewhere with similar types of independent front suspension, and while I - purely as a matter of personal preference - like to be kept rather more closely informed of what the front wheels are doing, there is no doubt that the Delage steering would be far less tiring on a long run than the more "live" variety, and also does everything that can be required of it.

Turning to the engine, it has eight cylinders in line with coil ignition, a downdraught carburettor and push-rod operated overhead valves. The dimensions of the cylinders are 80 mm. x 107 mm., giving an overall capacity of 4,302 cc and an R.A.C. rating of 32 h.p. Only 120 b.h.p. is claimed in the maker's catalogue, but in view of the weight (35 cwt) and the performance I should think this is conservative, and the figure of 142 b.h.p. recently quoted in *Motor* sounds more probable. This would be equivalent to the respectable amount of 33 b.h.p. per litre.

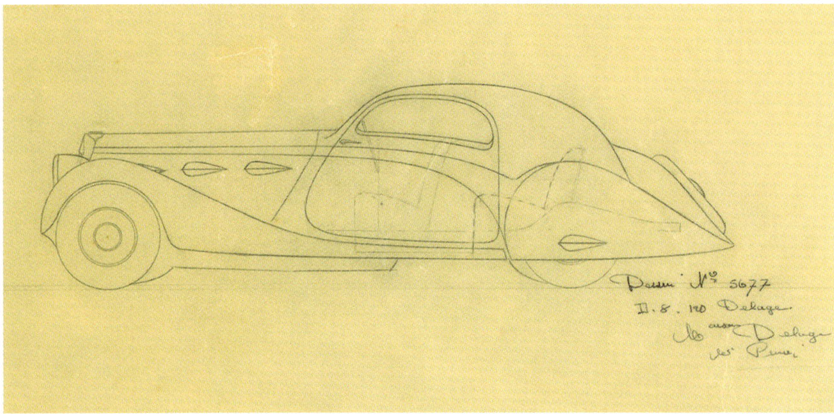

The five-bearing crankshaft seems to be absolutely rigid as there is no trace of period or vibration at any point in the whole range. I do not know what the recommended maximum revs. are, but about 4,500 seems to be comfortable and gives maxima in the gears of roughly 35, 48, 76 and 106. Why, in view of this, the calibrations on the rev. counter begin to assume a most menacing expression at 3,800 and disappear altogether at 4,200 is not entirely clear to me!

The performance is vastly assisted by the Cotal electrical "cog-swopping" arrangements, and to my mind this device is the only serious improvement on a good crash box with a clutch stop. At least two other types of easy-change gearboxes now popular are practically as large and as heavy as the rest of the car put together, and waste a shocking amount of power overcoming incidental friction. The Cotal box, on the other hand, is small, light and efficient and has given absolutely trouble free service in the hands of countless private owners.

This particular one, however, was temporarily out of adjustment owing to a dirty contact, which was apt to give a delayed action when changing gear. This, nevertheless, is quite an exceptional occurrence.

To a lazy driver, the Cotal box is just pleasantly foolproof, while to the connoisseur it is full of new possibilities. It is operated by a tiny lever within easy reach of the fingers of the left hand without moving it off the rim of the steering wheel. Ease of selection is assisted by the lever being spring loaded towards the lefthand side of the gate, where top and third are situated. The actual direction of the car, fore or aft, is decided by a short central lever of conventional type that is placed well under the dash and does not get in the way. When this is in neutral the Cotal box can be left in gear. The ratios are fairly close, but in view of the adequate power put out at moderate revs. there is no need for them to be any closer. On the other hand, it struck me that, for really fast getaways, bottom might well be a little lower.

The fear that a dead battery would make the Cotal box rather a white elephant is not really very serious, since, if there is enough current to start and run the engine, there will certainly be enough to provide the 3 1/2 amps. necessary to operate the ratios. It will be remembered, incidentally, that the Delage, which ran so impressively at Le Mans last year, was fitted with a Cotal box.

Upward and downward changes are usually made in the normal manner, speeding up the engine or allowing it to slow down while the gear is in neutral, before engaging the next one. Given even the most moderate judgement, there is then no need whatever to use the clutch. Alternatively, lightning snap changes can be made, both up and down, keeping the throttle open throughout, but in this case it is essential that the clutch should be used to take up the drive. Changes of this kind neither should nor need be done as a habit!

As to actual performance, this is really astonishing, considering the very considerable weight and frontal area of the car, while the Coachcraft body makes no very sensational concessions to considerations of streamlining.

On this actual machine the speedometer is 5 per cent. fast at the top end, and it has been timed at 105 m.p.h. with the speedometer showing 110. On my run it never passed 102 (97-98 actual) but I had stupidly left a window open which, of course, completely upset the airflow. There can be no doubt, however, that it is a genuine 100 miles-an-hour conveyance, with a little more in hand besides.

Acceleration up to 85 m.p.h. is really rapid, and only after that does one have to "wait for it" to any extend. Eighty-five is about the best cruising speed, at which the car will waft along in almost complete silence, at a very moderate throttle opening. As this only requires an engine speed of 2,600 r.p.m. there should certainly be no reason why it should not be able to maintain it indefinitely, while with such a moderate stroke the piston speed is quite reasonable.

The following acceleration figures can only be taken as approximate, since they were read with an ordinary pocket watch and by uncorrected speedometer readings. As against this, the gear box being out of adjustment was no possible sort of help so that, all things considered, they are probably not far from the truth. At worst, they show the very marked address with which speed is accumulated.

0 to 50 m.p.h. took 11 seconds, 0 to 70 took 17 1/2 seconds and 30 to 70 in third throughout, 12 seconds. The car will run smoothly at (and pull away from less than) 10 m.p.h. in top gear.

The brakes were badly out of adjustment and no figures were taken, but judging by the smaller D6-70 model, which I have also driven, they should be very powerful when in proper order. It is gratifying that a proper handbrake is provided; with modern traffic conditions one is more essential today than ever before.

The Marchal starter sets the engine going from stone cold in two revolutions at most.

Page 108 Delage D8-120 design study number 5677 by Letourneur & Marchand.

Page 109 1936 Delage D8-120 Aérosport, Chassis 51012, by Letourneur & Marchand. The first Aérosport built. James Patterson Collection.

Above left Chrome hub cap and wheel cover with the Delage name. It covers the spare tire and gives the car a formal look.

Right The Delage blue oval came to be associated with speed, quality and elegance. Chassis 51760.

Pages 112-113 Sales material for a Delage D8 faux cabriolet. It is offered with five places on a standard chassis and four places on a sports chassis.

DELAGE

The Coachcraft body was free from rattles and the wide doors still closed easily. The hood can be raised and lowered by one person in a few seconds and when up does not drum at any speed. Ingress and egress are easy and the driving position excellent, while after the rather lurid advertisement pictures I was surprised to find that both mud-guards were visible, so that I had no difficulty in placing the car at speed within two or three inches of the near-side kerb. The front passenger's seat is lower and does not provide such good support for the small of the back. The rear seats are exceedingly well contoured, but are too high off the floor, so that the legs of an average person are left dangling in mid-air. The back passengers feel more road shocks than the people in front, though even so they still ride in more than usual comfort, and much could doubtless be done by careful adjustment of the shock-absorbers. There is ample elbow room, both in front and at the back.

Under the bonnet the engine itself has all the expected Delage neatness, but in the change-over of bodywork the electric wiring had got moved about and presented rather a confused appearance. On this particular car the exhaust is taken through outside pipes and, owing to a somewhat faulty layout (provided, as I understand, against M. Delage's better judgement to special requirements, in the days of the original streamline body), a peculiar and not very pleasant beat is introduced into the otherwise unobtrusive exhaust note, when accelerating hard. This would not be perpetuated on ordinary models.

The front axle assembly is quite usual for this type of suspension - i.e. independent transverse leaf spring - and the castor angle is maintained by radius rods. The rear suspension is by normal semi-elliptic springs. The price of the model is tested at £1,370, which is certainly good value for money.

The D8-120: A Road Test

32 h.p. Delage D8-120 Coupé

English-built coachwork is fitted to the Delage, as now sold in this country, and the combination is an attractive one with a chassis that possesses good performance and handling qualities. The D8-120 is a straight-eight motor of 4.3-litre capacity. It makes a somewhat deceptive machine. Though geared high, it is flexible on top gear and can be driven leisurely without the driver realizing at first how much performance is in reserve.

It will amble about in town and along byways in an entirely docile manner. There is the power to keep pulling heartily against main road gradients, even if they are approached fairly slowly, without a change down, and steeper hills, such as the usual 1 in 6 1/2 test gradient, can be taken fast on third. Top gear acceleration is good, especially in the middle and upper range, the engine really "coming to life" at about 2,000 r.p.m., which is around 40 m.p.h. on top gear.

Particularly if the engine is run up on the lower ratios - a practice encouraged by the almost instantaneous changing made possible by the Cotal electro-magnetically controlled gearbox - the acceleration is of a high order, speeds between 70 and 80 m.p.h. being attainable on the comparatively limited stretches of clear road.

That the ultimate maximum is really useful is shown by the timed speed figures. Outstanding, however, in the running of this Delage is the way in which it maintains about 60 m.p.h. not feeling that it is hurrying in the least. A striking feature during the tests on Brooklands track was that as the engine became really warmed up maximum speed steadily improved, not the reverse, and although some fourteen miles were covered "flat-out" oil temperature did not rise above 90 deg. C. The normal road figure on this car is 40 degrees or less, due to the provision of a separate oil radiator.

DELAGE D8 Faux Cabriolet
5 places sur le châssis Normal — 4 places sur le châssis Sport

The engine is smooth, but not quiet under acceleration and there is some pinging when picking up from the lower speeds. Probably the engine would be quieter if the carburettor intake was fitted with an air silencer. A fairly pronounced deep exhaust note is produced, and sounds unusual today though actually it is not unpleasing. It is noticed chiefly at the lower speeds and on the gears, and at low speeds there is an illusive irregularity in the note.

High-geared, Accurate Steering

This car handles very well indeed. Its steering is high geared, needing only 2 1/4 turns from lock to lock, and is accurate and light, whilst the springs allow no side roll. A corner can be taken with delightful certainty, the driver placing the car just where he wishes. Although, then, the transverse-leaf-spring independent springing is not an overly soft system, it also provides excellent riding comfort, practically devoid of any pitching tendency.

The hydraulically operated brakes require fairly considerable pedal pressure for full results, but they are definitely brakes that can be put on hard when the car is traveling fast without upsetting the steering, and when an emergency stop comes to be made they give really good power. The handbrake lever is convenient on the right, and holds the car securely on a steep hill.

Now as to the Cotal gearbox. On the left of the steering column is a finger lever working in a visible miniature gate. This is in effect a switch, and is not pre-selective, the lever being moved at the moment of requiring a change, as with an ordinary gear lever. But the clutch pedal does not have to be used, being required only for starting. Control in traffic becomes a considerably easier process than usual.

When changing, the throttle pedal should be treated exactly as applies to an ordinary gear change, and with some practice a smooth take-up of the drive can be secured after a change has been made, even from top to second. The rev. counter fitted is a considerable help, for the engine soon gets up to the 4,000 r.p.m. mark, where warning lines are placed on the dial, though it will readily go beyond this range on the gears without appearing to be unduly forced. The gears are very reasonably quiet. The little gear lever is spring-loaded towards the third and topside of the gate, but there is no actual safety provision to prevent the two lowest ratios in the epicyclic type of gearbox being put in at a road speed too high for them.

Temporary neutral is obtainable by moving the finger lever into the center of the gate, as desirable when waiting in traffic, the clutch pedal not being light to hold down for any length of time. When the car is finally stopped the master lever rising from the floor is put into neutral, this also moving fore and aft for the forward and reverse gears respectively. It is rather far away from the left hand. The front compartment is left clear for getting in and out at either side.

In spite of the long bonnet, the driver feels able to handle the car confidently at first acquaintance. A big spring-spoked steering wheel is set well down, and the driving seat is adjustable for angle by a Leveroll mechanism, with the slide runners as fixed, the seat would not lock sufficiently near to the wheel to suit an average height driver. The whole adjustment could easily be moved forward, and then the position becomes good, except that the front of the seat cushion touches one's legs.

This drop-head foursome coupé by Coachcraft has proportions that admirably suit the chassis size. It is a handsome and most impressive car, the effect being added to by flexible exhaust pipes coming out of the bonnet.

Leather upholstery of excellent quality is used, and the seats prove very comfortable. At the back there are deep foot wells; a central folding armrest is provided. The rear seats are somewhat "blind" in this body, a remark that also applies to some extent to driving, as regards rear quarter vision when joining

another road, for instance. A practical position of this drophead body is with the front part of the head rolled back and neatly secured, giving a good open effect, the side window being usable to prevent draft. It is fairly easy to put the head entirely down, and when closed it has no particular tendency to flap.

Slow-reading Speedometer

A comprehensive range of instruments includes not only the rev. counter, but also water and oil thermometers. The speedometer proved to be slightly slow at 30, only 0.5 m.p.h. fast at 50, and slow again at 70, whilst the highest reading shown was 96.

There is good luggage space in the tail, and pleasing use is made here of cabinetwork - as well as in the interior of the car - for the lid of a neat tool tray.

The engine starts readily and will pull at once, seeming sometimes to be on the rich side when starting. It is a clean unit of impressive size, with accessible sparking plugs and oil filler. As is more often than not the case, there is no reserve petrol tap, though the tank is of large capacity. One-shot chassis lubrication is fitted.

32 H.P. Delage D8-120 Coupé

PRICE: with two-door-head foursome coupé body, £1370. Tax, £24
RATING: 31.6 h.p., eight cylinders, ohv, 80 x 107 mm., 4,302 cc
WEIGHT: without passengers, 38 cwt. 0 qr. 14 lb.
LB.PER CC: 0.99
TYRE SIZE: 700 x 17 in. on bolt-on wire wheels
LIGHTING SET: 12 volt. Non-variable charging rate.
TANK CAPACITY: 24 gallons; approx. normal fuel consumption, 14-15 m.p.g.
TURNING CIRCLE: (R) 46 ft. 5 in.; (L) 47 ft. 6in.
GROUND CLEARANCE: 9 in.

ACCELERATION

Overall gear ratios:	10 to 30	20 to 40	30 to 50
3.91 to 1	10.4 sec.	10.0 sec.	9.0 sec.
5.43 to 1	7.0 sec.	6.7 sec.	6.3 sec.
8.50 to 1	4.3 sec.	4.9 sec.	-
11.80 to 1	3.8 sec.	-	-

From rest to 30 m.p.h. through gears	5.3 sec.
To 50 m.p.h. through gears	12.6 sec.
To 60 m.p.h. through gears	17.6 sec.
To 70 m.p.h. through gears	24.4 sec.
25 yards of 1 in 5 gradient from rest	5.1 sec.

SPEED

	m.p.h.
Mean maximum timed speed over 1/4 mile	92.55
Best timed speed over 1/4 mile	97.83
Speeds attainable on the indirect gears (normal and maximum)	
1st	25-31
2nd	38-43
3rd	60-75
Speed from rest up 1 in 5 Test Hill (on 1st gear)	22.88

Brake Test: Mean stopping distance from 30 m.p.h. (dry concrete), 31.5 ft.
Weather: Dry, warm, bright, wind light, N.W. Barometer: 30.40 in.
Performance figures for acceleration and maximum speed are the means of several runs in opposite directions with two up.

Appendix

A Specification Checklist of Delage Racing Cars

Type	Year	Cyls	Bore X Stroke mm	Displacement cc	bhp @ rpm	Designer	Wheel Track Ins	Base Ins	No. made	Notes
Dieppe	1908	1	100 x 150	1178	17@ 1800	Causan	43.5	102.0	1	Plus two cars with two-cylinder de Dion engines.
X	1911	4	80 x 149.5	3006	50@ 2500	Michelat	55.9	115.8	4	Two horizontal valves per cylinder. Five-speed gearbox.
Y	1913	4	105 x 180	6234	118@ 2300	Michelat	F-54.0 R-51.0	108.0	3	Four horizontal valves per cylinder. Five-speed gearbox.
S	1914	4	94 x 160	4441		Michelat	53.1	107.8	3	DOHC, desmodromic. Five speeds, direct on third.
5L	1922	6	85 x 150	5107	170@ 3600	Toutée			1	Pushrod Bugatti-type radiator.
6L	1922	6	95 x 140	5954	185	Planchon			1	DOHC.
DH	1922	12	90 x 140	10,688	280@ 3200	Planchon			1	LSR car. Pushrod.
2LCV	1923	12	51.4 x 80	1992	116@ 6000	Planchon	F-51.1 R-47.3	103.9	1	Bugatti-type radiator. Four-speed gearbox. DOHC. Two valves per cylinder. Unsupercharged.
2LCV	1924	12	51.3 x 80	1984	120@ 6000	Planchon	F-51.1 R-47.3	103.9	3	Modified by Lory. Unsupercharged. Four-speed gearbox.
2LCV	1925	12	51.5 x 80	1973	205@ 6500	Planchon	F-51.1 R-47.3	103.9		More Lory modifications. Two Rootes blowers. Five-speed gearbox. Total five 2LCV built.
1.5L	1926	8	55.8 x 76	1486	170@ 8000	Lory	51.1	98.0	3	Two Rootes blowers in parallel at side of crankcase. Planchon based.
1.5L	1927	8	55.8 x 76	1486	170@ 8000	Lory	51.1	98.0		Same cars modified. One Rootes blower at front of engine. New cylinder block & exhaust system.
3L	1937-39	6	83.7 x 90.5	2988	137@ 5100	Engine Michelat based	F-53.8 R-57.5	125.5	2	Watney Le Mans cars. Three carburetors. Cotal gearbox. Delahaye chassis.
3L	1945	6	83.7 x 90.5	2988	142@ 5300	Engine Michelat based			6	Watney Le Mans cars. Three carburetors. Cotal gearbox. Delahaye chassis.

Speed Records All With Flying Start

Year	Location	Class/Distance/Duration	Driver	mph	Car
1924	Arpajon	World Unlimited - 1 mile	Thomas	143.31	DH
1925	Linas-Montlhéry Arpajon	World Unlimited - 5 mile International Class D (2001 - 3000 cc) - 1 mile	Benoist Divo	133.00 133.78	2LCV
1929	Brooklands	International Class A (over 8 litres) - 5 mile - 10 mile	 Cobb Cobb	 128.14 128.01	 DH DH
1930	Brooklands	International Class A (over 8 litres) - 5 mile - 10 mile	 Cobb Cobb	 129.36 129.29	 DH DH
1932	Montlhéry	International Class C (3001 - 5000 cc) 500km 3 hours 1000km 6 Hours 2000km 12 hours	 Eyston Eldridge Denly Don	 117.704 117.734 116.903 117.024 115.998 111.938	 D8 Spl. D8 Spl. D8 Spl. D8 Spl. D8 Spl. D8 Spl.

Year	Event	Driver	Place	Car Type
1906	Coupe des Voiturettes de l'Auto	Menard/Lucas	2nd	A
1907	Coupe des Voiturettes de l'Auto	Bonnard	7th	
1908	GP des Voiturettes de l'ACF	Guyot/Reyrol	1st	Causan
		Thomas/Houlet	5th	
		Lucas/Chenard	12th	
1911	Coupe de l'Auto des Voitures Légères	Bablot	1st	X
		Thomas	3rd	X
		Guyot	4th	X
1913	GP de l'ACF, Amiens	Bablot	4th	Y
		Guyot	5th	Y
	GP de France, Le Mans	Bablot	1st	Y
		Guyot	2nd	Y
		Duray	5th	Y
1914	Indianapolis 500	Thomas	1st	Y
		Guyot	3rd	Y
	GP de l'ACF, Lyons	Duray	8th	S
	Sioux City Track (300 miles)	Knipper	6th	Y
	Elgin Road Race (301 miles)	Knipper	3rd	Y
	Kalamazoo Track (100 miles)	Knipper	3rd	Y
1915	Chicago Speedway (500 miles)	L. Chevrolet	7th	Y
	Chicago Speedway (100 miles)	Oldfield	4th	Y
	Minneapolis Track (500 miles)	Oldfield	8th	Y
	Sheepshead Bay Speedway, Astor Cup (350 miles)	Limberg	6th	Y
		Oldfield	dnf	Y
1916	Indianapolis 500	Oldfield	5th	S
	Sheepshead Bay Speedway, Astor Cup (250 miles)	Le Cain	4th	S
		Devigne	7th	S
	Sheepshead Bay Coney Island Cup	Limberg	3rd	S
	Sheepshead Bay Queens Cup	Limberg	4th	S
	Sheepshead Bay Metropolitan Trophy	Devigne	2nd	S
	Sheepshead Bay Fifty-Mile Special	Le Cain	1st	S
1917	Sheepshead Bay Ten-Mile Record	Le Cain	97.8 mph	S
1922	Mont Ventoux Hill Climb	Thomas	1st	5L
1923	Vingt-Quatre Heures du Mans	Belben/Torchy	13th	DE
	GP de l'ACF, Tours	Thomas	dnf	2LCV
	GP of San Sebastian	Belben	2nd	6L
	Gaillon Hill Climb	Thomas	1st	DH
		Thomas	2nd	6L
		Thomas	3rd	5L
	La Turbie Hill Climb	Thomas	1st	
	Mont Ventoux Hill Climb	Thomas	1st	5L
	Argenteuil Hill Climb	Divo	1st	5L
1924	GP of Europe, Lyons	Divo/Fretet	2nd	2LCV
		Benoist/Carra	3rd	2LCV
		Thomas/Lhermitte	6th	2LCV
	GP of San Sebastian	Morel	3rd	2LCV
		Divo	4th	2LCV
	La Turbie Hill Climb	Divo	1st	2LCV
	Mont Ventoux Hill Climb	Divo	1st	2LCV
1925	GP of Europe, Spa-Francorchamps	Four entries	dnf	2LCV
	GP de l'ACF, Montlhéry	Benoist/Divo	1st	2LCV

The Delage Record In Competition

Year	Event	Driver	Place	Car Type
1925	GP de l'ACF, Montlhéry	Wagner/Torchy	2nd	2LCV
	GP of Spain, San Sebastian	Divo	1st	2LCV
		Benoist	2nd	2LCV
		Thomas	3rd	2LCV
	Gaillon Hill Climb	Thomas	1st	DH
	La Turbie Hill Climb	Benoist	1st	
	Mont Ventoux Hill Climb	Divo	1st	6L
1926	Targa Florio			
	(team withdrawn due to Masetti's fatal crash)			2LCV
	GP of Europe, San Sebastian	Morel	3rd	1.5L
	GP of Spain, San Sebastian	Wagner/Benoist	3rd	1.5L
	British GP, Brooklands	Wagner/Sénéchal	1st	1.5L
		Benoist/Dubonnet	3rd	1.5L
	GP de la Baule (France)	Wagner	1st	1.5L
	Gaillon Hill Climb	Benoist	2nd	2LCV
1927	GP de l'Ouverture, Montlhéry	Benoist	1st	1.5L
	GP de l'ACF, Montlhéry	Benoist	1st	1.5L
		Bourlier	2nd	1.5L
		Morel	3rd	1.5L
	GP of Spain, San Sebastian	Benoist	1st	1.5L
		Bourlier	3rd	1.5L
	GP of Europe, Monza	Benoist	1st	1.5L
	British GP, Brooklands	Benoist	1st	1.5L
		Bourlier	2nd	1.5L
		Divo	3rd	1.5L
	500 Miles of Rafaela, Argentina	J. Malcolm	1st	2LCV
1928	JCC 200-Mile Race, Brooklands	M. Campbell	1st	2LCV
	Trophé National, Boulogne	M. Campbell	1st	2LCV
1929	Indianapolis 500	Chiron	7th	1.5L
1931	Brooklands 1500 cc Lap Record (127.05 mph)	Howe		1.5L
	Brooklands Standing Start Lap-Record (115.55 mph)	Cobb		DH
1932	Avus, Germany, 1500 cc category	Howe	1st	1.5L
	Coppa Acerbo, Italy, GP category	Howe	1st	1.5L
1933	Nürburgring, 1500 cc category	Howe	1st	1.5L
1935	Albi GP, France	Howe	3rd	1.5L
	Prix de Berne, Switzerland	Howe	3rd	1.5L
1936	Brooklands, RAC Light Car Race	Seaman	1st	1.5L
	Brooklands, JCC 200-Mile Race	Seaman	1st	1.5L
	Donington 200 Miles	Seaman	1st	1.5L
	Isle of Man, RAC 1500 cc Race	Seaman	1st	1.5L
	Pascara, Italy, 1500 cc Race	Seaman	1st	1.5L
	Prix de Berne	Seaman	1st	1.5L
1937	Vingt-Quatre Heures du Mans	Gérard/De Valence	4th	3L
1938	Vingt-Quatre Heures du Mans	Gérard/De Valence	dnf	3L
1939	Vingt-Quatre Heures du Mans	Gérard/Monneret	2nd	3L
1947	Ulster Trophy, Ballyclare	Woodall	1st	1.5L
1949	Vingt-Quatre Heures du Mans	Louveau/Jovet	2nd	3L

First published in *The Delage Journal*, No. 29, Winter 1991/92, Great Britain.
Reprinted here by permission.

Year & Chassis Nos.	Type	Capacity in cc	Bore/Stroke	Cylinders	Valves	Wheelbase in meters	Notes
1905	A	1059	106 x 120	1	Auto		De Dion engine. Shown at 8th Salon de l'Automobile, Paris, December.
	B	496	84 x 90	1	Auto	2.05	
1906 (Nos. up to 120)	[A]						Factory moved to Levallois.
	[B]						
	D	697	90 x 110	1	Auto	2.05	De Dion engine Type C. 2nd place Coupe des Voiturettes.
1907 (Nos. 121 to 260)	[B]						
	[D]						
	E	939	100 x 120	1	Auto	2.05	De Dion engine.
	F	936	100 x 120	1	Auto	2.05	De Dion engine.
1908 (Nos. 261 to 639)	[B]						
	[D]						
	[E]						
	[F]						
	G	1201	80 x 120	2	Side		De Dion engine.
	H	1763	75 x 100	4	Side		De Dion engine.
	GP	1430	100 x 160	1	Side		28 B.H.P. at 2800 rpm First GP des Voiturettes de l'ACF at Dieppe (Guyot with Causan designed single cylinder. Thomas & Lucas with De Dion Twins).
1909 (Nos. 640 to 1000)	[D]						
	[E]						
	F	1018	100 x 130	1	Side	2.45	
	J	2116	75 x 120	4	Side		Ballot engine.
	L	1328	62 x 110	4	Side		Chapuis Dornier engine.
	M	1458	65 x 110	4	Side		Longer wheelbase and wider track than 'L'; Chapuis Dornier engine.
	N	1580	65 x 120	4	Side		Delage engine in 'M' chassis.
1910 (Nos. 1001 to 1400)	[D]						
	[F]						
	J	2116	75 x 120	4	Side		De Dion engine. Chassis price £290.
	M	1365	66 x 100	4	Side		De Dion engine.
	N	1843	70 x 120	4	Side		Delage engine.
	R	1456	65 x 110	4	Side		4G Ballot engine. Chassis price £238.
	T	1328	62 x 110	4	Side		4G Ballot engine. Chassis price £220.
	AB	2116	75 x 120	4	Side	2.85	4F Ballot engine. Chassis price £290.
	AC	1456	65 x 110	4	Side		"M" with Ballot engine.
	AD	2116	75 x 120	4	Side		"AB" with De Dion engine.
	AE	1365	66 x 110	4	Side		De Dion engine.

Year & Chassis Nos.	Type	Capacity in cc	Bore/Stroke	Cylinders	Valves	Wheelbase in meters	Notes
1911 (Nos. 1401 to 1980)	[R]						Chassis price £260.
	[T]						Chassis price £230.
	[AB]						Chassis price £290.
							Now with 4 speed gearbox and 4FM Ballot engine.
	TR	1328	65 x 110	4	Side		4GM Ballot engine.
							Chassis price £238.
	U	950	90 x 150	1	Side		De Dion engine.
	X	2983	80 x 149.5	4	Horizontal		Coupe de l'Auto; 4 cars with 5 speed gearboxes and 4 valves per cylinder horizontal.
1912 (Nos. 1981 to 4000)	[R]						New Courbevoie factory opens.
	[T]						
	[AB]						Chassis price £295.
	[TR]						Chassis price £238.
	AH	2562	66 x 125	6	Side	2.77	Delage designed engine in "AB" chassis. Chassis price £400.
	AI	2292	75 x 130	4	Side	2.90	
1913 (Nos. 4001 to 5050 and 10,000 to 10,250)	[R]						Chassis price £221.
	[T]						
	[AB]						Chassis price £292.
	[AH]						Chassis price £340.
	[AI]						Now available with long wheelbase. Chassis price £318.
	AK	2662	66 x 130	6	Side	3.36	Chassis price £358.
	Y	6234	104 x 180	4	Horizontal		4th at GP de l'ACF at Amiens; 1st Le Mans; 1st 1914 Indianapolis.
1914 (Nos. 5051 to 5950 and 10,251 to 10,400)	[AI]						Chassis price £308.
	[AK]						Chassis price £360. Now available with long wheelbase.
	AM	1590	65 x 120	4	Side	3.20	
	BI	2292	75 x 130	4	Side	2.95	
	BK	2260	65 x 130	6	Side	3.36	
	S	4430	94 x 160	4	O.H.V.		Twin Desmodromic valves; F.W.B.; 5-speed gearbox. 6th GP de l'ACF at Lyon.
1915 (Nos. 5951 to 6000)	[BI]						
	[BK]						
	CO	4524	80 x 150	6	Side	3.68	Prototypes only. Monoblock engine.
1916 (Nos. 6001 to 6255)	[BI]						Chassis price £406.
	[BK]						Chassis price £480.
							Detachable wheels for army use.

Year & Chassis Nos.	Type	Capacity in cc	Bore/Stroke	Cylinders	Valves	Wheelbase in meters	Notes
1917 (Nos. 6256 to 6755)	[BI] [BK] [CO]						100 chassis for French army.
1918 (Nos. 6756 to 6840)	[BI]						Detachable disc wheels standardized.
	[BK]						Detachable disc wheels standardized.
	[CO]						Detachable wire wheels standardized.
1919 (Nos. 6841 upwards)	[CO]						
	DO	3005	80 x 150	4	Side	3.43	Similar to CO but 4 cylinders. F.W.B. optional.
1920	[CO]						Chassis price £1000.
	[DO]						
	[GS]	4524	80 x 150	6	Side	3.43	A CO engine in the shorter DO chassis.
1921	[CO]						
	[DO]						Chassis price £1025.
	CO2	4524	80 x 150	6	O.H.V.	3.67	Similar to CO but with pushrod engine and 1/2 elliptic rear springs.
	GS						Engine now as CO2.
1922	[CO]						Chassis price £1100; complete car £1550.
	[DO]						Chassis price £775; complete car £1125.
	[GS]						Chassis price £1200; complete car £1600.
	[CO2]						
	DE	2116	72 x 130	4	Side	3.18	Chassis price £550; complete car £750.
	Sprint	5100	85 x 150	6	O.H.V.		Delage II or "La Torpille".
1923	[CO2]						
	[GS]						
	[DE]					2.98	"Sport" model.
						3.18	"Normal" model.
						3.43	"Long" model.
	Grand Prix	1992	51 x 80	V.12	O.H.V.	2.60	4 O.H.C. One car entered at Tours (retired).
	Sprint	10688	90 x 140	V.12	O.H.V.		World Land Speed Record 143.3 m.p.h. 24th July.

Year & Chassis Nos.	Type	Capacity in cc	Bore/Stroke	Cylinders	Valves	Wheelbase in meters	Notes
1924	[GS]						
	[CO2]						Also available with 3.43 meter wheelbase.
	[Grand Prix]						2nd, 3rd, 6th at GP de l'ACF at Lyons; 3rd at St. Sebastian
	DI	2120	75 x 120	4	O.H.V.	3.18	Chassis price £375.
	DIS					2.98	Engine details as for DI.
1925	[DI]					3.18	
	DIC					3.18	"Colonial" model. Wider track, otherwise as DI.
	DISS					2.98	
	GL	5984		6	O.H.V.	3.62	"Normal" model.
						3.85	"Sleeping" model.
						3.40	"Sport" model: also had narrower track. Also "Grand Sport" model available.
	[Grand Prix]						Now supercharged. 1st, 2nd at Grand Prix de l'ACF; 1st, 2nd, 3rd at Saint Sebastian.
	Sprint	5954	95 x 140	6	O.H.V.		Delage I. The new "La Torpille" with twin O.H.C.
1926	[DI]						
	[DISS]						
	[GL]						
	Grand Prix	1484	56 x 78	8	O.H.V.	2.5	Twin O.H.C. supercharged. 160 B.H.P. 5-speed gearbox. Exhaust on right-hand side. 1st & 3rd at British GP Fastest lap European GP.
1927	[DI]						
	DIS					2.98	Not to be confused with the 1924 DIS!
	[GL]						Sport model discontinued.
	Grand Prix	1484	56 x 78	8	O.H.V.	2.5	Induction and exhaust reversed. B.H.P. increased. 1st, 2nd, & 3rd French GP at Montlhéry; 1st & 3rd Spanish GP at Saint Sebastian; 1st European GP at Monza; 1st, 2nd, & 3rd at British GP.
	DM	3180	75 x 120	6	O.H.V.	3.50	
	DMS	3180	75 x 120	6	O.H.V.	3.21	Closer ratio gearbox than DM, also slightly narrower track.

Year & Chassis Nos.	Type	Capacity in cc	Bore/Stroke	Cylinders	Valves	Wheelbase in meters	Notes
1928	[DI]						
	[DIS]						
	[GL]						
	[DM]						
	[DMS]						
	DR	2516	70 x 109	6	Side	3.22	Basically the DMS chassis.
1929	[DM]						
	[DR]						
1930	[DR]						
	DRC						"Colonial" model.
							Wider track otherwise as DR.
	[DM]						
	DMN					3.39	Same track as DMS;
							otherwise as DM.
	D8	4060	77 x 109	8	O.H.V.	3.31	"Court" model.
						3.44	"Normal" model.
						3.64	"Long" model.
1931	[DR]						
	[DRC]						
	[DM]						
	[DMN]						
	[D8]						
1932	[D8]						
	D8S						Narrower track than D8;
							otherwise as D8 Court.
	D6	3045	77 x 109	6	O.H.V.	3.11	"Normal" model.
						3.25	"Long" model.
	DS	2516	70 x 109	6	O.H.V.	3.11	"Normal" model.
						3.25	"Long" model.
1933	[D8]						
	[D8S]						
	[D6]						
	[DS]						
	D6-11	2000	75 x 75.5	6	O.H.V.	3.06	"Normal" and "Sport" models.
						3.26	"Long" model.
1934	[D6-11]						
	D8-15	2660	75 x 75.5	8	O.H.V.	3.26	"Normal" and "Sport" models.
						3.53	"Long" model.
	D4	1498	77 x 79.5	4	O.H.V.	2.80	
1935	[D4]						
	D6-65	2650	79 x 90.5	6	O.H.V.	3.14	
	D8-85	3550	79 x 90.5	8	O.H.V.	3.34	
	D8-105	3550	79 x 90.5	8	O.H.V.	3.26	Narrower track than D8-85.

Passenger/Racing Car Production and Specifications 1905-1949

Year & Chassis Nos.	Type	Capacity in cc	Bore/Stroke	Cylinders	Valves	Wheelbase in inches	Notes
1936	D6-60 D6-80	2335 3227	74 x 90.5 80 x 107	6 6		125.5 133.4	Michelat engine. Based on Delahaye 18 CV.
1936-37	D8-100	4303	80 x 107	8		133.4 144.4	
1936-38	DI-12	2151	80 x 107	4			Based on Delahaye 12 CV.
1937	D6-60 D8-120	2528 4303	77 x 90.5 80 x 107	6 8		125.5 133.4	Michelat-based engine. New tax rating.
1937-38	D6-70 D8-100	2729 4303	80 x 90.5 80 x 107	6 8		125.5 144.5	Michelat-based engine. New tax rating.
1938	D6-80 D8-120	3553 4744 4490	84 x 107 84 x 107 75 x 84.7	6 8 V12		133.4 133.4 107.9	Based on Delahaye 20 CV. Prototype.
1939-40	DI-50 D6-75 D8-100 D8-120	2372 2798 4744 4744	84 x 107 81 x 90.5 84 x 107 84 x 107	4 6 8 8		117.7 125.5 144.5 133.4	Based on Delahaye 134G. Michelat based.
1946-54	D6 3L	2988	83.7 x 90.5	6		125.5 131.5	Michelat based. Only SWB available in 1948-49.

Brackets are used when information for that model has been included in previous years.

First published in *The Delage Journal*, No. 14, Autumn 1975, Great Britain. Reprinted here by permission.

Delage Chassis Numbers

1919	6840	7040
1920	7041	7850
1921	7851	8550
1922	8551	10300
1923	10301	13000
1924	13001	15650
1925	15651	18839
1926	18840	21600
1927	21601	25500
1928	25501	29400
1929	29401	32800
1930	32801	34150
1931	34151	35250

1932	35251	36550
1933	36551	37150
1934	37151	38400
1935	38401	38750
1936	38751	39854
1937	50000	51200
1938	51201	51550
1939	51551	51900
1940	51901	60200
1940-1946	WWII	
1946-1952	880000	880500

First published in *The Delage Journal*, No. 19, Autumn 1980, Great Britain. Reprinted here by permission.

Other historical documents may contradict the Delage chassis numbers shown in the table above. The Delage Company and the coachbuilders occasionally used chassis numbers out of sequence, particularly for show cars.

The Straight 8 Delages, 1930 - 1939

A Chronological Account Compiled by F.R. Hovell

1930 The first D8 Delage came on the market with a 4,060 cc o.h.v. engine, fitted to three chassis lengths:
D8 "S" or "C" with a wheel base of 10′ 10″
D8 "N" with a wheel base of 11′ 3″
D8 "L" with a wheel base of 11′ 11″

The engine was 77 mm. bore and 109 mm. stroke and had a RAC rating of 29.4 h.p.

The chassis prices for the three chassis lengths were £655, £650 and £655 respectively and tyre size on the two shorter chassis was 675 x 30 and on the "L" 675 x 32.

1931 The specification remained the same except for the addition of reaction arms to the front axle and the alteration of the tyre sized to 700 x 18 on the two shorter chassis and 700 x 20 on the "L". But chassis prices were now increased to £715 for the "S" and "N" and £725 for the "L".

1932 The Super Sports or Grand Sport version of the D8 was now on sale at a chassis price of £975. This was a sporting version of the D8 with a wheelbase of 10′ 3″. At this time the tyre size was standardised at 700 x 18 and the prices of the three "ordinary" models remained the same as in 1931.

1933 The main changes this year were in price. However the wheel base of the Grand Sport went up to 10′ 10″ - the same as the "S" model. A few of the "L" chassis had a modified rear spring arrangement. One thing which should be made clear is that the carburettor and manifold were water-heated and not oil heated as is sometimes stated.

1934/5 A completely new D8, the D8-15 was now available in two chassis lengths, giving wheelbases of 10′ 8″ and 11′ 7″.

The engine of the D8-15 was much smaller than the original D8: it was 2,668 cc with a bore of 75 mm. and 75.5 mm. stroke. The RAC rating of this, Delage's most square engine, was 27.9 h.p. - no concession to the horse power tax there. Tyre sizes on both chassis were 650 x 18, the front suspension was independent with a transverse spring and wishbones, each wheel having its own steering rod, with the track rod running across the chassis near the flywheel. This suspension was also used on the D6-11 and D4 models and was also used under license by Studebaker.

The valve gear of the D8-15 engine was of the divorced valve spring design and worked very well, though it was not easy to service.
The 4 litre D8 continued to be available during this period and some of the Grand Sport engines had this type of valve gear fitted. It was claimed to eliminate valve bounce.

Other new features of the D8-15 were Bendix type cable operated brakes and bolt-on wheels. Also, as a sign of the times, the cylinder blocks were no longer detachable.

1935 During this year came yet another eight cylinder car, the D8-85. This was rated at 31.1 h.p., had a bore of 79.2 mm. and a stroke of 90.5 mm., giving a capacity of 3,591 cc. Wheelbase was 11′ 0″ and chassis price £820. A long chassis model was also listed at £850. Tyre size on both chassis was 650 x 17.

The engine of the D8-85 reverted to normal push-rod and rocker valve gear. The carburettor was by Stromberg. For the first time on an eight cylinder car hydraulic brakes were fitted, with twin master cylinders. A striking feature was the clutch which had a large coil spring round the input shaft. As with all the Delage models from this time on the car had I.F.S.

1936 Yet another new model, the D8-100, was produced. The bore and stroke of the new engine were 80 mm. x 107 mm., giving a capacity of 4,300 cc and an RAC rating of 31.7 h.p. The wheelbase went up again to 11′ 1″, I.F.S. continued and carburetion was now by Solex. This was the first eight cylinder model to be fitted with the Cotal gearbox. Chassis price was around £790, including, of course, the outside exhausts with chrome covers.

1937 The wheelbase of the D8-100 was now up to 11′ 11″ but this model was joined by the "Super Sports" version called the D8-120. This had a wheelbase of 10′ 10″ and the price was around £875. Tyre size on both models was 700 x 17.

1938 The D8-100 continued unchanged but almost inevitably the wheelbase of the D8-120 went up to 11′ 1″.

1939 This year saw the appearance of the last of the great eight cylinder models to bear the name of Delage. This was the D8-120 with the 4,743 cc engine. Bore and stroke were 84 mm. x 107 mm. giving an R.A.C. rating of 34.99 h.p. Wheelbase was back to 10′ 10″, the Cotal gearbox and I.F.S. were of course continued. The brakes were now, however, of the self adjusting type and a Carter carburettor replaced the Solex. The D8 Series of cars have always lent themselves to the fitting of exotic coachwork and this was carried out by most of the great French Coachbuilders of the 30s. A few cars were to be found with English bodywork and some had shells by French coachbuilders finished and trimmed in England.

Reprinted here by permission. This article was written by the late Fred R. Hovell, a well-known Delage expert. It was first published in *The Delage Journal*, No. 3, Autumn 1968.

Tuning Specifications for the Delage D8

Engine:
Eight cylinder in line. O.H.V. push rod operated. Two valves per cylinder, in detachable cast-iron head. Separate detachable cylinder block.
Bore and stoke 77 x 109 mm. Capacity 4,060 cc
Crankshaft carried in 5 main bearings.
Nelson Bohnalite bi-metal pistons having aluminium crowns with lightweight cast-iron skirts.
High pressure forced feed lubrication, with drilled connecting rods for small end lubrication.

Ignition:
Delco-Remy coil and distributor.

Carburetion:
Single Smith-Barraquand 5-jet carburettor, heated by circulating engine oil.
D8-SS-100: 4 S.U.s on special manifold
D8-120: Dual choke Stromberg

Clutch:
Single dry plate.

Gearbox:
In unit with engine. 4-speed and reverse sliding pinion.
Overall ratios - Type D8-S and SS: 3-6, 4-89, 7-56, and 12-6 to 1.
Many variations of above to suit other chassis and coachwork with final drive ratios varying between 3-9 to 5-4 for normal D8 chassis.
D8-120: Cotal electro-magnetic 4-speeds and reverse.

Suspension:
Half-elliptic springs front and rear. Shock absorbers:
Type D8, various; Type D8-S and D8-SS, André Hartford tele-hydraulic; Type D8-15, D8-85 and D8-120, by transverse leaf springs and wishbones.

Frame:
Channel steel braced by channel and tubular cross members. Dropped frame on D8-S and D8-SS. The D8-15, D8-85 and D8-120 have a boxed steel frame.

Wheelbase:
Type D8, 10 ft. 10 in. to 11 ft. 11in.; Type D8-S and D8-SS, 10 ft. 2 in. and 10 ft. 10 in.; Type D8-120, 11 ft. 0 in.

Track:
4 ft. 8 in.

Wheels:

Type D8 steel disc with 18 x 7.00 tires; Type D8-S and D8-SS; Rudge Whitworth center-lock wire wheels with 18 x 7.00 tires.

Weight:

D8-SS chassis, 2,900 lb.

Performance data:

D8. Grand Sport chassis (Press Road tests).
Mean Time flying 1/2-mile, 98.9 mph
Maximum speeds on gears: 1st, 29 mph; 2nd, 48 mph; 3rd, 67 mph
Top in excess of 100 mph under not very favourable conditions.

Chassis data:

Valve clearances: inlet and exhaust 5/10 mm. or .018 in. (hot)
Contact breaker points: .018 in.
Firing order: 1, 3, 7, 4, 8, 6, 2, 5.
Carburettor: Single Smith-Barraquand 5-jet carburettor,
or Type D8-SS-100, 4 S.U.s on special manifold.
Plug Setting: 7 mm. or .023 in. Champion type No. 7.
Valve Timing: Type D8. Inlet: Opens 5° b.t.d.c. Closes 35° a.b.d.c.
Exhaust: Opens 45° b.b.d.c. Closes 5° a.t.d.c.
Types S and SS. -vary as to camshaft fitted.
Ignition timing: At t.d.c. fully retarded.
Distributor points: .018 in. gap
Sparking plugs: Champion type 7. Points setting .023 in.
Tracking: Front wheels, toe-in 3 mm. (measured at wheel centre height).

Select Bibliography

Automobile Quarterly. Second Quarter, 1976.Volume XIV, Number 2. New York.

Buckley, J. R. ***The Delage D8 Series.*** Leatherhead: Profile Publications, Ltd., 1966.

Dolfus, Alain. ***Delage.*** Paris: Editions Presse Audiovisuel, 1984.

Emmons, Harold H. ***The French Classics. 1925 - World War II. The Mad Years.*** Torque, January-February, 1983.

Gary, Pierre. ***Delage D8-S 1932 - "La belle voiture française".*** Autopassion, No. 35, May 1990.

Jolly, François. ***Delage. D8 100 et D8 120.*** Chanteloup-les-Viges: Association des Amis de Delage, 1996.

Jolly, François. ***La grande aventure des Delage 3 litres.*** Pontoise: Edijac, 1988.

Jolly, François. ***Delage - La D6-70 et ses evolutions.*** Cegedim: Association des Amis de Delage.

L'automobiliste. No. 66, September 1985.

Rousseau, Jacques. ***Cent ans d'automobile française.*** Paris: Flammarion, 1984.

Rousseau, Jacques. ***Guide de l'automobile française.*** Paris: Editions Solar, 1988.

Rousseau, Jacques. ***Les automobiles Delage.*** Editions Lariviere, Collection Fanauto, 1978.

Yvelin, Paul, Rousseau, Jacques, Charbonneaux, Philippe. ***Delage. Une belle voiture française.*** Chanteloup-les-Viges: Association des Amis de Delage, 2003.

Yvelin, Paul. ***Delage - Notes et Souvenirs.*** Levallois-Courbevoie.

Photograph Credits

Adatto Archive	Pages 10-11, 12, 13, 14, 15, 36, 37, 40-41, 42-43, 44 (bottom), 49, 58-59, 61, 64, 70 (bottom), 72, 76-77, 78, 80, 86 (bottom), 88, 89, 92-93, 95, 100-101 (top), 105, 108 (top) and 112-113.
Blackhawk Collection	Pages 84, 85 (bottom), 90 and 91.
Caron, Jean-Paul	Pages 18-19, 20, 21, 22-23, 24, 26, 28-29, 38-39, 45, 46-47, 48, 60, 62-63, 65, 73, 74-75, 79 (bottom), 85 (top), 86-87 (top), 87 (bottom), 96, 97 (bottom), 98, 100 (bottom), 104, and 106-107.
Figoni, Claude	Pages 79 (top), 81 and 82-83.
Furman, Michael	Back cover, Pages 6-7, 44 (top), 66-67, 68, 69, 70-71 (top), 71 (bottom), 74-75 and 117.
Rosenthal, Maurice-Louis	Page 33.
Rousseau, Jacques	Pages 16-17, 25, 27 and 32.
Stephen Photography	Pages 97 (top), 99, 108-109, 114-115 and 118-119.
Williams, Scott	Front cover, 8-9, 50, 51, 52-53, 54-55, 56, 57, 102-103, 110 and 111.
Wood, Thomas	Pages 30-31 and 121.
Zumbrunn, Michael	Pages 34-35.

Design	Ben Gibbs - Motion Design
Printer	Star Standard Industries Pte Ltd., Singapore
Printing	Four color litho on Roland Speedmaster
Page size	215 mm x 285 mm
Text paper	170 gsm Stora Enso Matt Art
End papers	140 gsm Woodfree
Dust jacket	150 gsm Glossy Art
Inks	Toyochem
Body text	9/13pt Frutiger 45 Light